ROUTLEDGE LIBRARY ]
URBAN PLANNI

Volume 15

# URBAN PLANNING IN A
# CAPITALIST SOCIETY

# URBAN PLANNING IN A CAPITALIST SOCIETY

GWYNETH KIRK

Routledge
Taylor & Francis Group

LONDON AND NEW YORK

First published in 1980 by Croom Helm Ltd

This edition first published in 2018
by Routledge
2 Park Square, Milton Park, Abingdon, Oxon OX14 4RN

and by Routledge
711 Third Avenue, New York, NY 10017

*Routledge is an imprint of the Taylor & Francis Group, an informa business*

*British Library Cataloguing in Publication Data*
A catalogue record for this book is available from the British Library

ISBN: 978-1-138-49611-8 (Set)
ISBN: 978-1-351-02214-9 (Set) (ebk)
ISBN: 978-1-138-48431-3 (Volume 15) (hbk)
ISBN: 978-1-138-48497-9 (Volume 15) (pbk)
ISBN: 978-1-351-05063-0 (Volume 15) (ebk)

**Publisher's Note**
The publisher has gone to great lengths to ensure the quality of this reprint but points out that some imperfections in the original copies may be apparent.

**Disclaimer**
The publisher has made every effort to trace copyright holders and would welcome correspondence from those they have been unable to trace.

# URBAN PLANNING IN A CAPITALIST SOCIETY

Gwyneth Kirk

Croom Helm London

© 1980 Gwyneth Kirk
Croom Helm Ltd, 2-10 St John's Road, London SW11

British Library Cataloguing in Publication Data

Kirk, Gwyneth
    Urban planning in a capitalist society. — (Social
    analysis).
    1. City planning — Great Britain
    I. Title II. Series
    333.7'7'0941      HT169.G7

ISBN 0-85664-929-5
ISBN 0-7099-0302-2 pbk

Printed and bound in Great Britain

# CONTENTS

# TABLES

# INTRODUCTION

In recent years British urban planning has attracted considerable criticism. There has been a boom in office-building, but many offices stand empty while thousands of people live in inadequate accommodation. New housing estates, whether in the suburbs or inner-city areas, often provide drab, monotonous environments, while some housing is uncompromisingly brutal. Factories, warehouses, docks and railway yards lie unused and deteriorating, apparently obsolete. Each year a little more good-quality agricultural land is taken up for suburban housing development. Traffic congestion gets worse, public transport more costly, and cross-town travel more time-wasting. These negative aspects constitute a severe indictment of the planning system. They are well documented, obvious to even casual observation, and borne out by everyday experience. Accordingly, I shall not detail them further. What I am concerned to explore are the reasons for some of these apparent contradictions. Why, for example, development companies are interested in building offices when so many people have difficulty finding decent housing at prices they can afford. Or why abandoned industrial sites and cleared housing areas in some inner-city neighbourhoods remain undeveloped, while productive farm land at the periphery of urban areas is built up, in the process of suburban expansion.

This book is about land-use planning[1] in a capitalist society, and considers several themes crucial to an understanding of the British planning system in particular. These include the role and significance of public intervention in a capitalist economy, especially as far as land-use planning is concerned, and consequent limitations to planning control; the role played by planning professionals, the majority of whom are employees of central and local government bureaucracies; opportunities for the public to influence land-use planning decision-making, and the scope for political action concerning planning and related issues of resource allocation. Land-use planning is both a technical and a political activity, concerned with the allocation of scarce resources — land and capital. Decisions as to how these resources are to be used have a serious effect on different sections of society, and may work to their benefit or disadvantage. Hence this book is about resource distribution and, ultimately, the structuring of power in British society.

The first chapter provides a preliminary discussion of land-use planning in Britain, and outlines the legal and institutional framework within which it operates. The emphasis is on the serious limitations to control over commercial development in a market-based economy, where land and capital are privately owned, and individuals and companies have the initiative in the development process. As a result, government cannot compel private development to occur.

Various different theoretical approaches have a bearing on land-use planning issues, and purport to provide explanations of them. These contributions come from the literature of academic sociology, geography, political studies and economics, from government policy and practice, and political activity. Four perspectives have been selected for detailed consideration on account of their influence and importance: the pluralist approach to the distribution of power, as derived from American community power studies, associated with such writers as Dahl (1961, 1967), Lipset (1960) and Polsby (1963); a bureaucratic approach, which stresses the importance of local (and perhaps to a lesser extent, central) government bureaucracy in affecting and managing people's lives and opportunities, associated with the work of Davies (1972), Dennis (1972) and Pahl (1975, Chs. 11 and 13); a reformist view associated with Blair (1973), Donnison and Eversley (1973), some of Harvey's essays (1973), and the urban programmes launched in the 1960s in the USA and Britain; work in the Marxist tradition, mainly following the structuralist approach of Castells (1976, 1977) and others. Chapter 2 is concerned with a summary account of these theoretical positions.

Three substantive areas emerge from my discussion of the British planning system which require more detailed consideration: the organisational context within which planning operates, the role of the professionals in their capacity as government officials, and the scope for public involvement and influence. These issues are taken up in the following three chapters, in connection with the theoretical approaches presented in Chapter 2. Chapters 3, 4 and 5 are thus concerned to elaborate on these three substantive areas in turn, and to see what light the different theoretical approaches can throw on them. The final chapter provides a concluding discussion of the strengths and weaknesses of the theoretical perspectives considered. Chapter 6 also draws out some of the implications of the arguments of the preceding chapters, and considers some possible developments and problems for planning in the future.

The book brings together various strands from a relatively wide literature, at present dispersed and fragmented within conventional boundaries of the different academic disciplines which have a bearing on land-use planning. This fragmentation is quite inappropriate, for planning straddles interdisciplinary boundaries, and makes a nonsense of them. To a remarkable degree 'social life', 'scarcity', 'power' and 'land use' have been, and are, conceptualised as distinct, separate, unrelated phenomena, the concern of sociology, economics, politics and geography respectively. This has had serious consequences for understanding, for there is thus a tendency only to focus on those issues which the particular academic discipline identifies as relevant. This led to urban sociology being reduced to studies of neighbourhood and community, friendship patterns and family life.[2] Urban geography has had a limited concern with patterns of land use, divorced from a consideration of the distribution of wealth and power in society.[3] Socio-economic questions, such as poverty, have sometimes been conceptualised as spatial questions, where the focus is on poor housing conditions and inadequate facilities, both in the home and the surrounding run-down neighbourhood. Poverty has also been redefined as a psychological and cultural issue, where the stress is on the personal inadequacy of poor people, their inability to cope with their situation, or to claim their rights from the various welfare agencies. But renewing the physical environment of the poor does not eliminate the causes of poverty, for this requires a substantial redistribution of resources.[4] Similarly, additional resources in the form of money, community workers and researchers directed to certain poor sections of cities are quite inadequate to tackle the problems of unemployment, low incomes and low educational achievement of people living in those areas.[5]

It is my hope that this book speaks to an area where the subject-matter of land-use planning and social science converge and overlap. It can be argued that there is a danger in this: that each topic will receive rather sketchy and superficial consideration, as it is obviously not possible to treat all the themes under review in depth within the scope of this one book. For this reason I have included a relatively wide bibliography so that readers can follow up such issues as interest them in the appropriate specialist works. Of course the bibliography is not exhaustive. Nor does it claim to cover what might be thought of as a 'complete range' of possible subject areas with a bearing on land-use planning, just those issues which seem to me to be the most significant. Thus I have tried to weigh this danger of superficiality and non-specialist treatment against what I hope is the usefulness of drawing together several themes and approaches usually found separately.

## Notes

1. The emphasis is on urban areas and urban planning, though land-use planning in Britain applies to rural as well as urban areas. I prefer the broader term 'land-use planning' to the 'town and country planning' in British practice or the 'city planning' in American usage. 'City planning' is too limited, and 'town and country planning' involves an unhelpful distinction between town and country, not easily defined. Ideally, the term to use should be 'planning', meaning both economic planning and its implications for land use, but clearly this is not appropriate in Britain where there is no national economic planning to speak of, and no national physical plan. In practice, 'planning' is used to stand for 'land-use planning', and I use it in this second sense on occasion.

2. A good example of this is provided by Carey and Mapes (1972). Also see Mellor (1977, especially Ch. 6) for a critical discussion of the weakness of 'urban' sociology.

3. This is discussed by Harvey (1973, Ch. 4).

4. Examples in Britain have been General Improvement Areas, Educational Priority Areas and Housing Action Areas. See, for example, Community Development Project (1977c); Halsey (1972); and Paris (1977).

5. As happened with the Community Development Projects of the late 1960s and early 1970s. See, for example, Chapman (1971) and reports published by the CDP teams.

# 1 URBAN PLANNING IN A CAPITALIST SOCIETY: SOME PRELIMINARY REMARKS

The purpose of this chapter is to outline the British land-use planning system, and it has three interrelated concerns: a discussion of the economic and political context within which land-use planning operates, the process of determining land-use priorities and planning objectives, and the institutional and legal provisions for controlling the commercial development process.

Land-use planning in Britain has its origins in the latter part of the nineteenth century, when public health legislation and building by-laws were introduced to check the worst excesses of the rapid, hap-hazard urbanisation which had taken place as part of industriali-sation. This historical development is well documented and need not be discussed here.[1] Over the years there has been increasing govern-ment intervention in the development process, as evidenced by the widening scope of town planning legislation.[2] The early Town Planning Acts made provision for the layout of suburbs, satellite towns and Garden Cities only, and though co-ordinated planning of the use of national resources was introduced during the First World War, it was seen as an emergency measure, and the administrative apparatus was disbanded in 1918. Government reports produced during the Second World War on the geographical distribution of the industrial population (HMSO, 1940), compensation and better-ment (HMSO, 1942b) and land utilisation in rural areas (HMSO, 1942a) all advocated a more systematic approach to the use of resources in peacetime, and have influenced subsequent legislation and policy to some extent.

The post-war period has seen large-scale reconstruction and redevelopment of war-damaged town centres and industrial areas, the establishment of 27 New Towns, the development of new shop-ping centres in both cities and suburbs, the building of offices, the creation of a national motorway network and through-traffic routes, new outer suburbs and the redevelopment and rehabilitation of inner-city housing areas.[3] Government policy at regional level has been concerned with inequalities between regions, and specifically with attempts to persuade firms to locate or relocate outside the relatively affluent South-East. The establishment of New Towns has been part of this dispersal policy, aimed both at reducing housing stress, particularly in London and Glasgow, and at providing new industrial infrastructure to stimulate local economies.

Thus there has been increasing public intervention in the use of land and buildings, but this intervention takes place within, and is circumscribed and limited by, a capitalist economic system. Since January 1948 all land in England and Wales has been subject to planning control, though in practice this has not been as sweeping as it might seem. Limitations to control over-development are many and serious, and rooted in the private ownership of land, private initiative in development, and the predominance of commercial interests, for commercial development in Britain is a business venture which, like any other, aims to make a profit.

The mechanism through which development control is exercised is the need to apply for planning permission. Proposals for new buildings, and the alteration or change of use of existing ones must be submitted for scrutiny and decision by local planning authorities whose job it is to prepare plans for their area.[4] The suitability of the proposed development for the site in question is considered with regard to neighbouring land uses, and to some extent with respect to the need for particular facilities in the locality. The planning authority will also have some idea whether there are likely to be other proposals for a specific location.

The over-riding justification for government intervention in the development process is that it is said to be in the public interest, the implication being that deference to private sectional interests is to be resisted. This rationale is usually formulated in terms of vague statements about 'safeguarding the environment', 'promoting high standards of design', 'conserving our civic heritage', 'using national resources efficiently', 'promoting suitable land use' and so on. Such statements are notoriously difficult to define, especially terms like 'efficiency' and 'suitability'. These formulations also raise problems for establishing priorities if objectives clash, or where resources are limited. It hardly needs stating that these aims may be open to question, variation in interpretation and emphasis, and potential disagreement and conflict between different individuals and organisations. Indeed, several distinct interests can be distinguished among the various 'actors' involved in the development process

— the planning authorities, landowners, contractors, development companies, financial institutions, industrial and commercial firms, conservationists, trade unionists, tenants and community organisations. This list can be subdivided further; for example, it is important to distinguish between fractions of capital. Financial institutions include banks, building societies which are mainly responsible for providing mortgage finance for owner-occupiers, and pension funds and insurance companies, which invest in the property market as a hedge against inflation. Some landowners, contractors and developers are interested in development, and can be distinguished from businesses which might be affected by it — either adversely or to their advantage. Many development companies are large national or international concerns, and can buy up land holdings to assemble large sites for comprehensive redevelopment. Large firms may also have their own property division, and may 'rationalise' their own plant and land holdings so as to capitalise on high-valued sites, or take over small firms, sometimes entirely for the sake of realising the value of the sites they occupy. The local planning authorities work within a context provided by higher tiers of authority, such as the Department of the Environment, responsible for co-ordination of land-use planning at national level. County plans provide a frame of reference for district authorities, and in London the Greater London Council is responsible for strategic metropolitan planning. Other planning bodies which aim to co-ordinate decisions of more than one local planning authority are the Joint Planning Boards for the Lake District and Peak District National Parks, and in London's Docklands the Joint Docklands Planning Committee.

Expansion of the scope of public intervention through land-use planning policies has been accompanied by an increase in professional staff employed by government — planners, architects, valuers, housing managers, urban researchers, public health officers and so on. Together with landowners, developers and financial institutions, they act as managers of the urban system, involved in deciding the location of industry, offices, housing, educational and health facilities and suchlike. Pahl (1975, Ch. 13), for example, points out the social nature of these distributive processes, for they are neither random nor arbitrary. This relationship between social structure and spatial structure is not entirely simple and straightforward, however, and it is important to consider it here.

### Social Structure and Spatial Structure

British society is characterised by inequalities of income, wealth, security., job opportunities and conditions, political power, educational opportunity, access to housing and so on (Butterworth and Weir, 1970; Giddens, 1973; Westergaard and Resler, 1975). Such inequalities have a spatial dimension and manifest themselves very noticeably in the differences in amenities and services available in residential areas. The different standards of housing, the incidence of gardens and parks, the kinds of shops serving a neighbourhood, the quality of schools, leisure and cultural facilities, the standard of treatment offered by doctors, the amount of traffic, parking space, noise, safe places for children to play, for instance, clearly vary from neighbourhood to neighbourhood, such that there are wide variations in desirability between localities. People are distributed throughout urban areas according to a variety of constraints placed upon them — most importantly, the availability of housing they can afford, and given this, their perceptions of what constitutes a convenient or attractive residential area (Gracey, 1969; Pahl, 1975, Parts 2 and 3; Simmie, 1974, Chs. 4 and 5). Further, as Rex and Moore (1967) have pointed out as a result of their study of Asians in Birmingham, people in the same position in the job market may have differential access to housing, depending upon their length of residence in an area, access to capital, skin colour and so on.

The spatial arrangement of land uses, as well as which uses are permitted and which are not, has important consequences, costs and benefits for different sections of the population. Examples are the cost and location of housing, which should be considered together with the cost and ease of transport, particularly for essential journeys to and from work. Services and amenities are clearly not evenly distributed, giving more or less equal access to all. Indeed they tend to be concentrated in some areas and absent from others, and while people may place different values on various facilities, convenience, accessibility and amenity seem to be generally valued.

Land-use planning, together with other instruments of social policy, may be expected to play an important role in redistributing spatial resources in favour of less well-off sections of society, though in practice it does not necessarily do this. As Pahl notes:

> In some cases the power position of a minority may be reinforced by planning decisions — as for example when local trade and business interests concentrate resources in city centres, raising

land values and banishing the poor to the periphery. The effect of this is to doubly penalise the poor: they get fewer rewards from the economic system anyway; now, in addition, they are further from urban facilities and so may have to pay more to get the same services as the rich (1975, p. 148).

Several writers stress the redistributive potential of the land-use planning system, and urge that policies based on compensatory principles be pursued. For example, Gans (1968) insists that planning should be 'user-oriented', and that planners should find out and take account of people's wants and needs rather than following their own preferences. Planning should be compensatory, and self-consciously redress the balance in resource distribution engendered by the workings of the market. Third, it should be 'for today', and cater for current needs, rather than investing a large proportion of resources in schemes designed to meet some assumed future demand or standard. Harvey (1973) is interested in social justice. This leads him to explore the interrelationship between social structure and spatial structure, and to consider the many detailed ways in which the poor are disadvantaged. He also examines the redistributive capacity of the planning system, and argues for the political will necessary to make the most of this theoretical potential. These discussions have as fundamental assumptions the important inequalities in income and wealth in Britain and America, and the possibility of reducing the effects of these differentials through the allocation and price of housing, ease of access to facilities, and so on. Pahl (1975, p. 257) supports this view, but argues that in the last analysis the quest for absolute spatial equality is futile, due to the unique character of every location, and the unevenness in distribution of physical and geographical features. He claims that 'without compensatory interventionist policies by the central administration even a society with an egalitarian wage structure would engender its distinctive spatial inequalities.'

There appear to be three different underlying principles which may guide public intervention in resource allocation: positive discrimination in favour of groups disadvantaged by the market mechanism; a reflective situation whereby planning follows the market and people's access to facilities is based on their ability to pay; or a reinforcing mechanism, where the planned distribution of resources discriminates in favour of those advantaged by the economic system, and hence widens the differentials which already exist. In Britain, all three principles are in operation. The provision of council housing is an important example of compensatory policy. Other examples are provided by the several positive discrimination programmes introduced in the late 1960s and 1970s: Educational Priority Areas (Halsey, 1972), which provided additional resources for schools in poor areas, General Improvement Area policy, whereby improvement grants were made available to property-owners to encourage and permit them to upgrade the condition of their property, Housing Action Areas (Paris, 1977), where the local authorities have been involved in acquiring and improving poor-quality housing. These programmes are compensatory in intention, if not always in effect. For example, the use of improvement grants often led to the displacement of the earlier tenants as landlords raised rents on their improved premises, despite the fact that they had not paid for the improvements entirely from their own pockets. An obvious example of resource distribution based on people's ability to pay is provided by privately owned housing, either owner-occupied or let by private landlords, where the price is fixed as a result of the relationship between supply and demand in a given locality. Examples of policies which reinforce the workings of the market include traffic management and environmental improve-ment schemes introduced into residential areas. These are designed to improve the quality of the immediate neighbourhood by restricting access to through traffic, limiting the amount of non-resident parking space, planting trees, and so on. At the same time, through traffic is diverted from these areas, and further congests the main roads surrounding them. Conservation area policies, intended to protect buildings of architectural or historic interest also reinforce the exclusive nature of such buildings, and enhance their price accordingly, placing them beyond the reach of all but the most wealthy section of society.

Over the past ten years or so there has been a mushrooming of local organisations with an interest in land-use planning engaged in trying to influence the future development of particular pieces of land, the rehabilitation or renewal of residential areas, the routeing of major roads and motorways, the location of large developments such as airports, power stations and so on. It is important to differ-entiate between the resources, aims and operational styles of these organisations, for they vary enormously from small-scale campaigns working on a shoestring to groups of people with considerable organisational skill, money and professional or political contacts.

Definitions are not straightforward here. What might be called 'working-class' organisations are generally campaigning for improvements in housing conditions, housing management, the provision of nurseries and play space and suchlike.[5] 'Middle-class' amenity groups are concerned with less basic issues, such as retaining and refurbishing old buildings, retaining old street patterns, enhancing the environment and introducing traffic management.

Generally speaking, there are two arenas within which negotiations over the distribution of resources take place: at work and at home. Improvements in pay and working conditions may be won as a result of bargaining with employers. Improvements in one's living environment may be achieved through the political process, either directly or indirectly. Examples are security of tenure provided by the Rent Acts, registering a 'fair rent' with the Rent Officer, or influencing local or central government on some aspect of housing or land-use planning policy. These two processes, taking place within the context of home and work, are interconnected, in that gains in wages and better conditions at work may be reinforced or eroded by costs and conditions at home, such as a rise in rent or rates, a deterioration in social and community facilities, an increase in noise, pollution or traffic congestion, and so forth.

In Britain at present there is relatively little direct trade union activity over 'place-based' spatial issues. The main union effort is concentrated on the interests of members in their workplaces, and concerned with raising incomes, shortening working hours, improving fringe benefits, reducing noise, dust, the likelihood of accidents, resisting redundancies and so on. However, there is some involvement, as for example in the rent strikes organised against increased council rents brought in by the Conservative's Housing Finance Act, 1972 (Sklair, 1975). Other examples include lobbying for retention of jobs that goes on at both central and local government levels if there is the suggestion of closure and redundancy, and the current campaigns against hospital closures brought about through cuts in public spending.[6] At a personal level there are links between people who are involved in their union at work, and who are also active in campaigns based in the neighbourhoods where they live.

The relationship between social structure and spatial structure is thus very important, and to some extent finds expression through political activity at both central and local government levels, in attempts to influence public intervention in the distribution of scarce resources.

I am concerned specifically with public intervention in connection with land-use planning, though clearly it has a much broader focus in contemporary Britain, incorporating all aspects of the welfare state, and including policies for maintaining and supplementing personal incomes, the provision of social facilities such as education and health care, and social services for the elderly, mentally and physically handicapped. It also involves policies aimed at maintaining employment, and job creation. Not all of the policies mentioned here have direct implications for land use, however.

The growth of public spending in the post-war period and the increasing intervention in what has otherwise been thought of as the domain of the private sector has led some writers to argue that there has been a taming of pure capitalism, and hence that it is a misnomer to use the term 'capitalist' at all. This debate is worth considering, because more than a difference of terminology is at issue. It involves a basic orientation to the economic and political arrangements of British society, and raises fundamental questions concerning the degree to which public intervention can be compensatory, and what, if any, are the limits to this.

### Public Intervention in a Market Economy

It is maintained by many people that Britain's economic and political system has changed qualitatively, at least since the Second World War, such that it is no longer accurate to think of it as a market economy. Hence the use of the terms 'mixed economy', 'post capitalism', 'neo-capitalism' which are preferred by such writers as Bell (1960), Burnham (1943), Crosland (1956, 1962), Dahrendorf (1959), Galbraith (1957, 1967), Strachey (1956) and Zweig (1961), and which have passed into everyday expression and conceptualisation. Crosland, for example, asking 'is this still capitalism?' answers in the negative, thus: 'almost all the basic characteristic features of traditional pre-1914 capitalism have been either greatly modified, or completely transformed... I believe that our present society is sufficiently defined, and distinct from classical capitalism, to require a different name' (1956, pp. 66, 68). There are several lines of argument developed at length by Crosland and others: that the worst social injustices of capitalism have been tamed by the establishment of the welfare state, full employment and the affluence of the 1950s (this latter held to be both widespread throughout the society and on the increase); that the general practice of appointing business managers has led to a separation between ownership and management of firms and an accompanying weakening of the power of owners; that there has been considerable and increasing state intervention, regulation and control of the economy and business decisions; that the power of capital has stimulated the development of strong labour organisations which constitute a 'countervailing power' to that of business managers and owners, and hence that there are no deep class antagonisms any more.

This perspective has been challenged forcefully in Britain by such writers as Blackburn (1967, 1972), Broadbent (1977), Frankel (1970), Miliband (1969) and Westergaard (1972) with reference to the 'rediscovery' and acknowledgement of the existence of poverty in the 1960s; the lack of evidence to suggest that there has been any major change in the inequality of wealth, income or educational opportunity comparing post-war with pre-war Britain; the lack of evidence to support the contention that managers do not run businesses in the interests of the owners; the continued prevalence of private ownership of land and capital, where a firm's decisions are taken on an individualistic basis and in the service of profit maximisation. In this context, labour organisations are always in a structurally weaker position to that of managers and owners. The wealth of evidence that these writers have mustered all supports the view that despite differences of degree on a number of points, Britain remains a capitalist economy, and hence it is a violation of the facts to call it anything else.

Blackburn (1972, p. 164), for example, argues that not only is the British economy capitalist, but that the modern form of capitalism with a widening role for the state as an enabler and facilitator of private enterprise, 'far from mitigating or abolishing the fundamental contradiction of capitalism, rather poses this contradiction *in a purer and more dramatic manner*' (stress in original). Interpretations of the mechanisms involved here are considered in various explanations of the current economic crisis.[7] Modifications to the market mechanism have been introduced through state intervention to enable the system to operate somewhat more rationally, and at less risk to individual firms. These include wage and price controls, taxation, licensing, the regulation of monopolies, state investment in firms, the underwriting of otherwise risky projects requiring very considerable capital outlay or lengthy research, guaranteeing a market for products, and a degree of capitalist economic planning where information is exchanged between the state and big private companies and their plans 'harmonised'. Further, there is the provision of services by publicly created bodies such as British Rail, the Port of London Authority, the Central Electricity Generating Board, the nationalisation of former private businesses, and state shareholding in private companies. The recent growth of this latter activity at local and regional level has been documented by Minns and Thornley (1978) and specifically in Coventry by Benington (1976). There are also provisions for partnership schemes between local authorities and private developers (HMSO, 1972b; Falk, 1974). In advanced capitalism the characteristic role of the state is that of both enabler and restrainer of the capitalist mode of production, with an increasing degree of penetration of the private sector. The state acts as a major employer, a purchaser and supplier, a transferer or subsidiser, a regulator, and is also concerned with steering the economy.[8]

The fact that considerable state intervention occurs is nowhere in dispute. What is contested is the significance of these trends, and the purposes and interests that are served by them. Put very simply, those who argue for a 'mixed economy' view see public intervention as a taming and diluting of capitalism, whereas those who argue for a 'capitalist economy' view see it as enabling capitalism to continue. A third perspective is the corporatist position, associated with Pahl (1977), Pahl and Winkler (1974) and Winkler (1977), where it is argued that Britain is developing a corporatist economy: 'an economic system in which the state directs and controls predominantly privately owned business according to four principles: unity, order, nationalism and success' (Winkler, 1977, p. 44). The role of the state changes from being facilitative towards and supportive of capitalism to being a directive one. The crucial distinction to be drawn is that between state influence on and state control of the internal decisions of private firms. Winkler accepts that Britain does not have a corporatist economy as yet, but argues that it is developing in this direction. Pahl is less cautious, and seriously overstates this case in my view, claiming *inter alia* that the state has been forced 'into the role of protecting and controlling the environment' and involved in the active protection of British industry as a result of the incursions of multinationals. However,

> this fundamental expansion of the state's role is not resisted by private capital; indeed paradoxically it is welcomed.

Private capital would doubtless have preferred a supportive role from the state but when a directive role came it was in no position to argue (1977, p. 162).

It is important to note the tense of this last statement, for according to Pahl this transformation from capitalist to corporatist economy has already happened. Westergaard (1977) gives a cogent critique of corporatism, organised around three themes: the sources of control of economic affairs, the criteria by which control is exercised, and the distributional outcome. He argues that the over-riding purpose of government controls over business is 'to restore profitability to private business in face of a persistent trend of falling profits' (p. 179). Second, even if state activity were directed against monopoly profit, where would the impetus for this come from? Clearly not from business, nor as things stand from labour. The British labour movement has shown 'more practical concern to run capitalism with greater efficiency — and some pay-off to labour — than to subvert capitalism rendering the principle of private profit inoperable' (p. 181). Third, it might be expected that a different economic order would have its distinctive allocation pattern, though Winkler does not suggest this. He acknowledges that a corporatist system would have the same inequalities as the capitalist mode of production, and the implication is that these would be based on the same sources and principles: property ownership for a few, and the need for the propertyless to hire out their labour power. Arguing against Pahl's and Winkler's view, Westergaard maintains that Britain does not have a distinctive corporatist economic system, marked off from capitalism. Furthermore, he sees no evidence of this developing in the near future. These authors simply fail to provide convincing arguments as to how the role of the state will change, and which groups in society have both an interest in bringing this about, and the ability to do so.

The government is involved in a balancing act, and continually needs to weigh the demands made by various competing interests, particularly sections of labour and business, so as to provide an economic and political climate within which business can operate profitably. Although certain kinds of intervention and regulation restrict the activities of firms, taken overall, public intervention in the economy is enabling a fundamentally capitalist system to continue to operate. Hence it should not seem paradoxical, as argued by Pahl, that some degree of state regulation and control is welcomed by private capital. The British government is working to support private business enterprise in a context of unfavourable competition from other national economies, particularly the USA, West Germany and Japan, demands for higher wages from organised sections of the labour force, and resistance to redundancies caused by structural changes in the economy and technological innovations such as automation. Broadbent, for example, notes that

> The UK economy is in decline relative to other Western industrial countries; it has a legacy of outdated industrial stock and outworn social capital. Yet the economy remains a basically capitalist market structure, dependent on profitability and the generation of surpluses for new investment to survive in a world of increasingly severe competition (1977, p. 28).

Accordingly, he emphasises the limitation of public intervention where 'the public sector operates *within* the market economy' (p. 8) and argues that this ultimately constrains the ámount of public intervention that can occur. Similarly, Westergaard and Resler conclude:

> Predominant power...lies with a ruling interest whose core is big business. The strength of business is manifest in its ties with a variety of other influential groups and bodies: directly with the Conservative Party, the commercial press and a range of pressure groups; less directly but none the less effectively with the machinery of state and the broadcasting media. These links are formed in part as bonds of common experience among top people. But the power of the ruling interest is founded in the set of common assumptions which govern the routine workings of economy, government and mass communications. Those assumptions — the core assumptions of the society — indicate the central place of business, because they are business assumptions: principles of property, profit and market dominance in the running of affairs (1975, pp. 275-6).

A capitalist mode of production involves an essential contradiction: a fundamental tension between the profit-maximising dictates of the economy, and the political desideratum that people's needs are provided for, as pointed out by Blackburn (1972), for example. The system is anarchic in two senses: first, the ultimate

goal of capitalist enterprise is capital accumulation and profit maximisation rather than the satisfaction of human need; second, the
mechanism of a capitalist mode of production — the market
system — is not subordinated to human control. Blackburn also
notes the unevenness of capitalist expansion, and the waste this often
entails. Industrial sectors are neglected or retarded, and a persistent
imbalance between public and private goods manifests itself. The
relationship between a capitalist economy and the natural environment is predominantly one of wasteful plunder in the pursuit of
private profit. Rivers are polluted, dust bowls created, air con-
taminated, and precious resources squandered. Heavily populated
regions, like North-East England and the north-eastern sector of the
USA, are condemned to stagnation and decay, as the economy
develops. Mellor (1977, Ch. 1) discusses uneven development
between regions in terms of dominance and underdevelopment,
concepts employed in the study of dominant cities or regions in
underdeveloped countries. Kapp (1978) gives a detailed account of
the social costs of business enterprise, stressing the waste of
resources, both natural and human. Friedman (1977) touches on
regional differences in his account of class struggles at work. He
analyses the relationships between firms, and between management
and workers in terms of the concepts 'centre' and 'periphery'. He
argues that different management strategies are aimed at different
groups of workers, with the objectives of allowing flexibility in the
firm's operation, and the maintenance of profit levels. Central
workers are those who have essential skills, or who contribute to
management's authority. They also include those who, by the
strength of their resistance to management, force managers to
regard them as essential. Firms in an industry may also be considered
central or peripheral, on the basis of their relative monopoly power,
as with small engineering firms, for example, 'squeezed between
large raw...materials suppliers (such as British Steel Corporation
and the oil companies) and large car and engineering firms' (p. 115).
Similarly, a pattern of differences within and between geographical
areas can be traced to these distinctions between central workers and
firms, and peripheral workers and firms, such that there is a per-
sistence of deprivation alongside areas of prosperity (Ch. 10).

Gough (1975) focuses on state expenditure, and discusses what he
calls the 'law of combined and uneven development', whereby not
all areas of a country, region or town attract development and
investment at the same rate. Some areas are more inviting to business

than others, partly depending on their historical development and current attributes in terms of population, markets, raw materials, communications, existing infrastructure and the quality and suitability of these for modern industrial and commercial activity. Also important is the comparative attractiveness of other locations for successful business enterprise. Areas which are expanding tend to attract further investment. They then require additional service provision, and thus draw additional investment and so on. This situation leads to a patchiness in investment and an imbalance between areas. In Britain this phenomenon is noticeable at the national level and at the city level.

At national level this unevenness is often referred to as 'the regional problem' and has been the subject of various reports and measures intended to induce firms away from the expanding South-East and Midlands to other regions, at present characterised by a legacy of nineteenth-century industrial infrastructure, mainly obsolete, as in Lancashire and Yorkshire, or marginal areas such as Cumbria, Northumberland and the Scottish Lowlands.[9] This uneven development has led to severe imbalances between the supply and demand for labour, housing, jobs and suchlike, and disparities in

**Table 1.1: Rates of Unemployment, UK, 1967-77 (percentages)**

|  |  |  | 1967 | 1971 | 1977 |
|---|---|---|---|---|---|
| UK |  |  | 2.3 | 3.4 | 6.2 |
|  | England |  | 2.0 | 3.0 | 5.8 |
|  |  | North | 3.9 | 5.7 | 8.4 |
|  |  | Yorks/Humberside | 1.9 | 3.8 | 5.8 |
|  |  | East Midlands | 1.6 | 2.9 | 5.1 |
|  |  | East Anglia | 2.0 | 3.1 | 5.4 |
|  |  | South-East | 1.6 | 2.0 | 4.5 |
|  |  | South-West | 2.5 | 3.4 | 6.9 |
|  |  | West Midlands | 1.8 | 2.9 | 5.8 |
|  |  | North-West | 2.3 | 3.9 | 7.5 |
|  | Wales |  | 4.0 | 4.7 | 8.1 |
|  | Scotland |  | 3.7 | 5.8 | 8.3 |
|  | Northern Ireland |  | 7.5 | 7.8 | 11.1 |

Source: *Annual Abstract of Statistics, 1979 edn (HMSO, London, 1978), p. 167.*

levels of amenities and service provision between regions.[10]

Despite public intervention in the economy, unemployment figures are rising, even though post-war governments have accepted the need to try to maintain full employment. During the economic expansion of the 1950s and early 1960s this was a relatively simple matter, though growth had clear regional effects. As the economy has become more depressed this issue has taken on increasing importance, both as far as numbers of unemployed people are concerned, and the marked regional variation in unemployment rates, as shown in Tables 1.1 and 1.2.

**Table 1.2: Unemployment Rates Amongst Economically Active Males, 1971: Selected Inner London Boroughs Compared to Development Areas**

|  | Population (thousands) | Unemployed (per cent) |
|---|---|---|
| *Best Development Areas* | | |
| Furness | 108 | 3.5 |
| Central Wales | 84 | 3.7 |
| Borders (Scotland) | 102 | 3.7 |
| *Mid-range Development Areas* | | |
| South-West Scotland | 150 | 5.7 |
| Falkirk/Stirling | 251 | 6.1 |
| *Worst Development Areas* | | |
| Tyneside conurbation | 805 | 9.5 |
| Central Clydeside conurbation | 1,727 | 11.0 |
| *Selected Inner London Boroughs* | | |
| Islington | 202 | 7.1 |
| Lambeth | 308 | 6.6 |
| Southwark | 262 | 8.2 |
| Tower Hamlets | 166 | 9.7 |

*Sources: London: The Future and You — Population and Employment (Greater London Council, London, 1973,) quoted in Falk and Martinos (1975, p. 5).*

From these regional statistics the South-East, including London, would appear to be least affected by unemployment, but averages always conceal extremes, and there are pockets of relatively high unemployment in parts of inner London, as shown in Table 1.2.

In the present context, public intervention is concerned with safeguarding jobs, granting subsidies to employers as an alternative to redundancy, and the establishment of Job Creation and training schemes. Together these have been estimated at keeping more than 900,000 people in employment over a three-year period.[11] Other possibilities currently under discussion involve various work-sharing arrangements, including early retirement, a shorter working week, longer holidays, or employing more people on a part-time basis to do what is currently being done by fewer people working full-time. The traditional land-use solution to the unemployment problem has been to try to attract industry to provide additional jobs. Indeed, the entire regional economic policy since the war has partly been directed towards this end, though with only mixed success, as the unemployment figures show,

Uneven development at the city level manifests itself in the growth of peripheral suburban areas, with new housing estates, light industrial sites, modern low-rise warehousing and storage space with good access to the motorway network. There has also been the growth of suburban shopping centres, catering for private car-users. In contrast, the inner-city areas are characterised by their largely Victorian infrastructure — old housing and school buildings, narrow roads, lack of open space and so on. Such neighbourhoods house a disproportionate number of elderly people, unskilled workers and claimants, and generally make greater demands on the social services than the outer suburbs. There is also a much higher incidence of multiple occupation and overcrowding in these areas.

Urbanisation comes about as a result of industrialisation, with a concentration of job opportunities in urban areas, the relative attractiveness of urban life as opposed to rural life, and the contemporaneous or subsequent mechanisation of agriculture, thus displacing erstwhile rural workers from the land. There is a need for housing, schools, shops, parks and social facilities of all kinds to provide for the needs of the urban work-force. Such services, however, are not profitable compared with investment in industry or agriculture, land development or speculation in commodities, and hence are unattractive to capitalist business enterprise. Castells sums up this contradiction as 'the need to concentrate labour power in industrial metropolises and the inability of capitals to provide an adequate supply of housing and facilities due to the lack of profitability' (1976, p. 166).

In the past the state has intervened in the housing market in Britain in order to alleviate severe housing stress. Direct control over workers' housing conditions includes building regulations first introduced in the latter half of the nineteenth century, and the subsequent provision of subsidised council housing. Intervention has also been indirect: regulating the relationship between landlord and tenant, as with the Rent Acts, formulating the rules under which building societies provide mortgages, and making public finance available to housing associations through the Housing Corporation, and so on.

Despite this intervention, however, the housing problem remains, seemingly endemic, and is acute in many inner-city areas. This is evidenced by the high demand for local authority housing relative to supply, the comparatively high cost of owner-occupied housing which is beyond many people's means, the condition of much inner-city housing, a substantial proportion of which either lacks facilities or is in need of repair, the degree of sharing and subsequent overcrowding in some areas, the decline of the private rented sector and the inadequacy of alternative provision for people who do not qualify for council housing and who cannot afford mortgage repayments, and the incidence of squatting (Allaun, 1972; Bailey, 1973; Berry, 1974; Donnison, 1967; Greve *et al.*, 1972; Murie *et al.*, 1976).[12]

Several factors seem to be responsible for this situation, either singly or in combination. An increase in the number of households and a decrease in household size has led to greater demand for housing units, partly met by increased suburban expansion and the division and conversion of larger housing units into several smaller ones. The demand for owner-occupied housing has taken many people away from the inner-city areas, which have a predominance of local authority tenures and privately rented accommodation. In any case, the availability of privately rented housing has declined drastically. Third, planning blight, due to uncertainty over urban renewal schemes, redevelopment for offices, urban motorways and road widening all add to the deterioration of particular premises as owners are unwilling to make improvements, and to a general rundown in the neighbourhood, decline in local services and so on. Obsolescence also plays its part. Many of the housing units in inner-city areas are in poor condition, though too much can be made of this point. It is also true that some parts of the inner-city have been greatly improved recently and 'gentrified',[13] and of course some inner-city areas have long been the home of the wealthy. A further point concerns the decline in manufacturing industry, much of which was located in inner-city areas. This imbalance between inner-city and suburban areas is acknowledged by governments to constitute a problem, hence the positive discrimination programme already mentioned, and more recent attempts to direct additional resources to inner-city areas (Community Development Project, 1977a, 1977c; Department of the Environment, 1977; Falk and Martinos, 1975).

There are two points I should like to make in connection with this stress on the capitalist mode of production: first, to suggest that it is worth noting the differences between capitalist countries, and the varying scope of state intervention in the economies of, for example, the USA, Japan and Western Europe. Hence I am not arguing that capitalist modes of production are entirely homogeneous, with Britain a 'typical representative'. As far as land-use planning is concerned, controls in Britain appear to be more far-reaching than they are in the USA, for example (Clawson and Hall, 1973; Hall, 1975, Ch. 9; Rock, 1972). Hall (1977) considers the planning of the 'world cities' of London, Paris, Randstad Holland, Rhine-Ruhr, New York and Tokyo and describes both differences and similarities in these very large-scale urban areas. He compares approaches to tackling planning problems concerning transport provision, decentralisation, housing form, the question of the continued dominance of central areas and so on. Hayward and Watson (1975) contrast the British, French and Italian experience of planning, and Friedmann (1974) provides a detailed account of differences between the USA, Britain, France, Italy, Nigeria and Turkey as far as the relationship between public and private enterprise is concerned in these varying mixed economies. Borja (1977) describes urbanisation in Spain, stressing the very weak controls over housing and inadequate provision of basic services in the 1950s, which has gradually given way to a more co-ordinated and interventionist policy. He also discusses the importance of workers' organisations in demanding improved housing at realistic rents. Lindberg *et al.* (1975) provide a varied collection of comparative studies of advanced capitalist societies, arranged around four central themes: planning *v.* the market, inequality *v.* opportunity, efficiency *v.* legitimacy, and dominance *v.* vulnerability. They acknowledge differences and similarities in historical circumstances, government policy, economic robustness and so forth, but stress that their central preoccupation is with capitalism, not industrialism.

However, it should be pointed out that some of the issues I consider are not necessarily specific to a capitalist mode of production. There are problems of housing provision and allocation, attraction of labour to isolated areas, planning urban and economic growth and suchlike, experienced in the centrally planned economies of Eastern Europe and the Soviet Union. This is described in general terms by Dobb (1970), for example, and by Pahl (1977), who stresses the importance of a comparative approach, where insights gained from a study of centrally planned societies can help towards an understanding of resource allocation in Western economies. Pahl also mentions the specific problem of attracting labour to expanding industrial areas in Siberia, where incentives of new housing and higher wages are offered in an attempt to stop the very high labour turnover these new settlements have experienced. Kovačerić (1972) and Zawadzki (1972) report on planning and government in Belgrade and Warsaw respectively, and also hint at difficulties, particularly that of in-migration to the cities which severely strains service provision, especially housing and public transport. The inference is that this in-migration has been virtually impossible to control, and has led to a considerable degree of unplanned growth in these cities. Konrad and Szelenyi (1977) describe difficulties of providing adequate housing in Hungary, and Musil (1968) writes of residential differentiation in Prague. Beckerman (1974, pp. 44-6) reports on high levels of pollution in industrial operations in various parts of Eastern Europe. Jeffery and Caldwell (1977) discuss general problems of resource allocation, writing about urbanisation in China.

Thus I am not arguing that only the capitalist mode of production involves problems of resource allocation. Any national government needs to provide a legal and institutional framework within which people can live and meet their needs, however these are defined. There have to be mechanisms for production and distribution, taking into account specialisation between geographical areas and groups of people. Channels of communication for expressing needs and mechanisms for co-ordination and decision-making are all required. In theory there is room for considerable variation in these arrangements along a number of dimensions. For simplicity these may be expressed as polar types, though in practice any number of different combinations are possible and may occur. Examples are the degree to which production and consumption are centrally planned and organised, or left to local or dispersed organisations;

whether or not the production of goods and services is in public or private hands, or what mix of these prevails; the degree to which services are provided universally, or on some selective basis; the importance attached to professionalisation and the role of experts; the dominance of high-cost, complex technology, or whether more simple technology is (sometimes) adequate. Any one of these dimensions can involve problems of resource allocation.

However, what I am arguing in the context of contemporary Britain is that the planning problems and limits to planning which exist arise out of the existence of a private market in land and capital. They are not simply attributable to general problems of production, co-ordination and resource distribution in large-scale industrial societies. Irrespective of the fact that there may be some apparently similar problems in cities in planned and market-based economies, I am arguing that it is the contradictions of British capitalism that account for the contradictions of land-use planning, and that the continuation of the capitalist mode of production will simply reproduce such problems. In Britain, to put it plainly, land tends to be put to its most profitable use, depending on the local circumstances, for the underlying criterion of commercial development is profitability, not the usefulness of particular buildings nor local people's need for them. For example, offices and shops involve a much higher return on investment than industrial space or housing, for firms are prepared to pay higher rents for office premises than most people can afford to pay for housing. Hence there will be some pressure to redevelop land for offices and shops or to convert other buildings to these uses, though there is a ceiling to this pressure in any one location, and a point at which the local market will not bear any more offices or shops. This means that development for land uses with a relatively low return on investment, as, for example, low-cost housing or open space, will not be provided by private capital as long as there are other outlets for investment. To the extent that these facilities are provided, this must be by state intervention of some kind, either direct or indirect. I shall return to this theme several times, for it is a fundamental one underlying the whole policy and practice of land-use planning in Britain.

Public intervention takes place through the established institutions of the political and administrative system. As far as land-use planning is concerned, it is important to note aspects of the British political and administrative context which have a bearing on its scope and operation, and through which land-use priorities and objectives are determined and implemented.

## The Political and Administrative Context of Land-use Planning in Britain

Every government since the Second World War has supported some degree of land-use planning. Hall (1972) presents the post-war Labour and Conservative records in this field, and shows that both are based on a consensus view of politics. Generally speaking, there are few clear lines of party division over planning policy, which is thought to be a matter of common sense, acceptable to men of goodwill. A notable exception to this is the 1951 Conservative government's denationalising development value, in order to encourage and promote development. Many local planning issues are decided by local planning authorities. Such protection of 'democracy' that the system of local government formally provides in areas like land-use planning where local authorities do have some discretion is not in operation, for it depends on the ventilation of issues through party conflict and debate, and this occurs only spasmodically and with little clarity of lines.

The methodology of plan formulation summed up by Geddes (1949, Ch. 16) in the rubric, 'survey, analysis, plan' is the basis of much planning work. If the survey is done properly, it is claimed, the facts of the situation will be incontrovertible and an objective analysis will enable a feasible, non-controversial plan to emerge. There is an acknowledgement that conflicts of interest exist between competing land uses, but planners tend not to recognise the validity of conflicts of interest between various groups in society. Since the political consensus is based on the proposition that those areas of policy on which all political parties agree are just common sense, and arise out of the nature of the situation, then this methodology has also wedded planning to a consensus view of politics. As Palmer (1972) notes, in adopting a rational problem-solving stance, the professional planner is free to maintain his neutrality, and at the same time deal with those issues which the consensus identifies as 'problems'. These have to be easily distinguishable and open to 'feasible' solutions. If the planner strays far outside this brief he calls into question his own neutrality. The success or failure, of planners' activity even within the system's own terms depends to a great extent on the way in which the problems manifest themselves. In order for the planner to act, it is not enough for a problem to exist, its existence must be recognised by those power groups for which he works, and

hence his assumptions and working context are implicit in the maintenance of the present structure.

In that land-use planning is concerned with the allocation of
scarce resources, it is a political activity. Every plan is a statement of
priorities, and a political document. One of the problems of judging
a plan is that the value content is often not spelled out, but implied.
If there is to be any satisfactory way of knowing the aims of a particular plan, and of ascertaining whether they have been achieved, or
are capable of achievement, a clear statement of objectives is necessary at the outset, sufficiently detailed and capable of measurement,
so that the future situation can be compared with the intentions of
the plan.[14] Such goal setting is properly the concern of politics, and
should be open to public discussion and influence. There are two
main problems with this. The first one concerns the ostensibly
technical nature of land-use planning, and the fact that it is not easily
accessible to 'non-experts', the second concerns the implicit goals of
a capitalist mode of production and the largely unquestioned power
of land ownership and development capital.

The first problem, involving the technical nature of land-use
planning, hinges on the very real problem of part-time, unpaid
elected councillors having effective control over their full-time,
salaried, professional officers. In theory it is the councillors who
take policy decisions, and the officers who organise the day-to-day
administration of their departments and advise councillors on
technical matters, though in practice this distinction between policy-
making and day-to-day administrative routine is a notoriously
difficult one to sustain at both local and central government levels
(Crossman, 1975, 1976, 1977; Fry, 1969; Hanson, 1969, pp. 91-103;
Hill, 1972; Miliband, 1969, pp. 47-9). The power of senior local
government officers largely depends upon their professional
expertise and their position in the organisational hierarchy. It is reinforced by the local authority structure itself, so that officers have a
near monopoly of information as compared to councillors, and the
more senior officers as compared to the more junior ones. Chief
Officers are in a position where they are usually allowed to define
what constitutes a problem, how it is conceptualised, what actions
might be taken to alleviate it, and what specific course of action they
recommend. Their recommendations are thus backed up by their
professional competence and authoritativeness. In principle, all
councillors have access to all the council's employees and all the
information in the local authority's keeping, should they wish to

refer to it. In practice, they tend to be given specific answers to particular questions, and being relatively unfamiliar with a department's administrative routine compared with the officers who work there full-time, councillors may not always ask the most pertinent questions. Committee chairmen have direct access to their Chief Officers and it is in the context of this working relationship that checks can operate if necessary, though even the chairmen are reliant on their officers for information. Accessibility to the public is a different matter, however. It is the more junior officers who have direct contact with members of the public and local organisations as a rule. Senior officers are generally protected from this, though they may appear at public meetings to help to put across the council's policies.

The problem of local councillors having effective control over their professional officers and retaining the right to take policy decisions themselves — in practice as well as in rhetoric — has been the subject of much comment and concern. It is also relevant to other local government departments, but the problem is particularly acute in the sphere of land-use planning due to the complexity of the subject-matter and the technical language and abstract diagrams which are used to express policy statements and to present proposals.[15]

Generally speaking, planners see themselves as adopting a rational, problem-solving stance, from a position of political neutrality. They define land-use planning as a technical activity, implying that planning matters are not the proper concern of the layman, and it is difficult to break through the mystification of planning policy and procedures. Technical and political matters do not bear such straightforward distinction. Nor are 'technical' matters as clear-cut as might be thought, as it is possible to argue many planning issues two ways. An example would be a proposal to develop a site which is not well served by transport facilities. The case against the application will rest on the fact of poor accessibility. On the other hand, it might be argued that new development will provide just that extra incentive necessary for increased investment in transport facilities to be worth while, and hence that poor accessibility should not be a stumbling block to development.

This is not to accept a 'professional conspiracy' view, however. It must be pointed out that such mystification is very often unconscious, with planners themselves often not recognising that there could be feasible alternatives. But whether consciously intended or not, it is true to say that to challenge proposals it is necessary to know something of the technical jargon and procedures, and to have access to adequate information. The severe handicap suffered by the general public in this respect extends to councillors who, as laymen, are very much beholden to their professional officers. Rather than being openly set by the elected representatives, planning goals often creep in by default, as a by-product of ostensibly technical standards of orthodox planning — density, zoning, notions of appropriate adjacent uses, physical fitness or obsolescence of buildings, appropriate minimum or maximum distances between different facilities, aesthetic considerations and so on. Thus such standards and values are largely set by planners, who to some extent dictate standards of consumption. In theory this is done by council members, who formally approve decisions.

Pressure for involvement in the planning process emanating from local organisations can be seen as a condemnation of councillors' ineffectiveness, and as an attempt to force local authorities to take account of people's views, and to make local government more democratic. Conventional constitutional theory saw no need for public involvement in land-use planning except to allow objections to be heard, because the authorities are elected. If the party system provided effective machinery for democratic participation, or if land-use planning was an apolitical process in which no clashes in policy were conceivable, that would make sense. But these premisses have been increasingly questioned.

Particular forms of direct communication between public and planners have been suggested in an attempt to get round the problems of representation by councillors. Indeed they would tend to render the councillors' role obsolete, though the claim that councillors represent the local electorate is already weakened by the low turnout at local elections. Hill (1970, p. 45), for example, states that under half the local electorate use their vote. The Committee on the Management of Local Government (HMSO, 1967a) reported that 42 per cent of the electorate in boroughs and urban districts voted in May 1964, 41 per cent in contested counties.

Where pressure for involvement comes from local groups, the councillors' role may be bypassed if planning committee members abdicate responsibility by recommending that the Borough Planner is the man who knows all the answers and should be contacted direct, as was the experience of the Millfield residents, for example, reported by Dennis (1972). Where pressure for direct contact with the public comes from planning officers, they themselves seek out 'user needs' rather than being instructed by council members. In recent years there has been an attempt by some planners to build larger and more powerful empires for themselves in the reorganisation of the management of local government, with its emphasis on providing co-ordinated comprehensive services. Eversley (1973), for example, advocated a much more broadly based managerial role for the planner of the future. Reynolds (1969, p. 132) went so far as to claim that the planners' task is 'to analyse and synthesize the goals and values of the community'. That this could be defined as a technical activity is quite remarkable, and has serious implications for the problem of answerability to the public for planning activity.

The second major difficulty with goal-setting is that of the general relationship between land-use planning and market forces, and the extent to which plans must conform to market principles to be capable of achievement, irrespective of whether investment is to be undertaken by the private or public sector. There are assumptions built in to plan preparation, and often unrecognised by planners, about the availability of resources and the interests to be taken into account. Business interests must not be antagonised. Assumptions are made about central government policy and the financial feasibility of development schemes, and so forth. It would be a sterile exercise, it is argued, to make plans which cannot be implemented within the prevailing economic and legal context, where development must be viable financially, and in most cases land acquired at market prices. This situation is at the very heart of the dilemma of land-use planning. In the words of a practitioner,[16] 'modern city building has become a helpless by-product of a gigantic investment bingo. Sites are developed and redeveloped for the sake of big return' (Amos, 1973). This means that there will be serious problems of providing low- and medium-cost housing, open space and social facilities, especially in areas of high demand for land, and hence inflated land values. For the shape and layout of the city is very much a three-dimensional reflection of the power structure in society, the outcome of the competition for the scarce resources of space, amenity and accessibility, and to recognise this is to recognise the severe constraints within which land-use planning operates, as Pahl (1975, Ch. 7) argues, for example. Also, and more fundamentally, business interests are taken into account automatically, as argued by Westergaard and Resler (1975), and not thought of as sectional. In fact 'the public interest' is not treated as though it diverges from business interests. In a capitalist economy it is assumed that what is to the advantage of business is, directly or indirectly, to the advantage of us all. Hence it is important to stress that the apolitical character of land-use planning as conventionally conceived is a nonsense.

This brings me to a consideration of the mechanics of land-use planning and details of planning procedure, with an emphasis on the problems involved in the continued private ownership of land and development values.

### The Commercial Development Process

For the developer, the purpose of the commercial development process is to increase the value of land by developing it for a particular use. The developer co-ordinates the various specialists and organises the different stages of a development project. He has personal and institutional links with estate agents, architects, financiers and contractors. Through them land is acquired, buildings designed and planning permission obtained, financial backing secured, building operations undertaken, and arrangements made with prospective purchasers or tenants. From each of these specialists the developer requires a commercial service. He is not normally overly concerned with either the aesthetics or the purpose of his buildings. The objective is to develop sites and to make a profit (Ambrose and Colenutt, 1975; Cadman and Austin-Crowe, 1978; Marriott, 1967).

Funding is mainly undertaken by the large financial institutions — banks, insurance companies and pension funds. Insurance companies, for example, have large funds to invest, and a recognition that it is more profitable to develop property directly rather than through investment in companies has led to a number of takeovers by insurance companies.[17] In other cases institutions have built up sizeable holdings in the development company, or secured a share of the equity, usually 10 per cent, in return for long-term credit. The amount of money invested in property by pension funds has also risen,[18] and there seems to be little difference between the investment strategies of public and private pension funds. Marriott (1967, p. 132) noted that the Co-operative Insurance Society, for example, owns 10 per cent of the equity of Oldham Estates, and comments that this 'particularly canny investor in property' illustrates a 'paradoxical link between an extreme manifestation of capitalism and an arm of the workers' movement'.

Confidence in property as a secure investment and a hedge against inflation is based on the unique nature of land and property, the fiscal advantage of income from property, and the growth of rents since the Second World War. There are three recognisable sub-markets in commercial property: industrial, shops and offices. To the developer and investor these sub-markets are mainly distinguished by their rate of return. Investors are concerned with financial security, income from investment and asset growth over the medium and long term. The growth potential of different types of property investment and property as compared to other investments is under constant review, though the property market in general has attracted increasing investment in recent years. The speculator is interested in property for capital gains, entering the market at periods of rapid growth, buying and quickly reselling at a profit. For him, property is a commodity like any other and his investment depends upon the booms and slumps of other commodities. The importance of the speculator's activity in the property market for users, developers and investors is that it can force prices up to a level where only the wealthiest can pay (Ambrose and Colenutt, 1975, Chs. 1 and 2; Colenutt, 1975).

The demand for offices has steadily increased, due to structural changes within the economy whereby an increasing number of people are employed in the non-manufacturing tertiary sector, and the expansion in office floor space per employee, partly arising out of the growing use of office machinery. Restrictions on supply in the form of government controls such as the need for Office Development Permits,[19] and the limited compass of the business area, for example the City of London, valued for the opportunity for face-to-face contact between businessmen, have resulted in successive rent rises. This general trend is expected to continue, and provision has been made for regular rent review clauses in most office leases. In general, property values have risen faster than the rate of inflation in recent years, whereas gilt-edged and government stocks selling at fixed rates of interest lose value with rising inflation; some shares have also lagged behind the inflation rate.

The explanation sometimes offered for this is that land is a commodity which has a fixed supply, while the demand for land and property is always expanding. The result barring a complete collapse of the economy is a steady increase in land prices, rents, and property assets. This is something of a circular argument because it is finance capital which is partly responsible for stimulating demand for property in the first place, by its own needs for office space, by lending money to developers and by pouring money into property as a hedge against inflation (Ambrose and Colenutt, 1975, pp. 44-5).

During the property boom, roughly between 1966 and 1974, in Britain, commercial rents increased, peaking in 1974. There has been a fall since then, though the London office market has picked up recently, helped by the advent of a Conservative government.

The urbanisation process has been accompanied by physical planning and the growth of infrastructural investment. The state increasingly undertakes development to complement that of private developers, providing roads, water supply, sewerage systems, public housing, industrial estates and so on. These services are characterised by their large-scale, collective nature, often requiring substantial capital to be invested before any return will be forthcoming. Thus whenever private development takes place, a whole range of public services must be provided in order for it to operate. This has led Lamarche (1976) to stress this facilitative function, and in his formulation the role of the state is reduced to serving the needs of private capital. Cullingworth (1972) makes this same point, but with a different emphasis, arguing that, increasingly, governments are using this social investment as part of physical planning or regional policy to help determine the location of economic activity, rather than passively reacting to it. Cullingworth sees this as evidence of the strengthening of the power of government in relation to private capital. While this may appear to be so in some specific circumstances, I am arguing that this reactive posture is fundamental to the system of planning and control of development in Britain. If government agencies are themselves landowners they can be more effective in controlling land use than where they are not, but they enter the land market on virtually the same terms as any other prospective buyers and generally have to pay the market price for a particular site.[20] State landowners would include the Ministry of Defence, Department of Education, Area Health Authorities, local authorities and nationalised industries such as British Rail, National Coal Board, and so on. Falk (1974) urges that local authorities undertake more development in their own right, and step beyond the confines of their traditionally accepted role, becoming more innovative rather than merely reacting to planning proposals submitted by private developers.

This brings me on to a consideration of the details of control over the development process, and the legal framework within which these controls operate.

**Controlling the Development Process**

Three aspects of land-use planning can be distinguished: plan formulation, and plan implementation comprising both development control, which can be thought of as negative planning, and public development, which can be thought of as positive planning. Plan implementation is based on the proposals, permitted uses and intentions of the development plan.

Each local planning authority is required to have some statutory plan for its area against which to judge proposals for development.[21] These plans have to be approved by the Minister, implying some minimal coherence and consistency at national level, although there is no overall national plan, a deficiency which has several important implications for local authorities, for the state as a whole, and for business interests.

There are two implications for local authorities. In the course of plan preparation, assumptions have to be made about national and regional policies and trends, and also assumptions about what is likely to happen in adjoining areas, but with very little factual information to go on. There is, hence, a need for hundreds of planners 'forecasting' likely future trends. Such assumptions may or may not be realistic or borne out by events because there are too many unknown or uncontrollable variables. Second, local authorities tend to be in competition with each other to attract high rate-paying land users, or business enterprises to enhance local employment opportunities. In less developed areas particularly, this may lead to competitive provision of infrastructural investment in an attempt to attract so-called 'footloose' industry, or one key development which, it is argued, will have the effect of attracting further development.

For the state, the lack of a national plan means that objectives like 'regional balance', implicit in the measures for containment and decentralisation referred to above, are unlikely to be met under the land-use planning system as it currently operates. Land-use objectives are given little consideration at national level, where the stress is on economic aims. The assumption is that the infrastructural side of economic activity — development or stagnation — will somehow sort itself out by a series of complementary readjustments where necessary. However, if there were such an overall physical plan, then it would be possible to apply criteria other than commercial viability to proposals for development. The need for jobs or housing in a particular area could then become a criterion for industrial location or residential development. Clearly, for the idea of a national plan to be accepted in principle, and for it to work in practice, the state would need more control over the location and activities of firms than it has at present. It would also need control over investment such that investors could not transfer finance to enterprises abroad even if the prospect of investment in Britain compared unfavourably. Thus the absence of a national plan works to the advantage of developers and businesses, allowing them to operate under only limited constraints. It is such people who have the most to lose, and who presumably would oppose the idea, were it proposed. In practice this has not been necessary for the suggestion has not been canvassed, exemplifying the power of routine assumptions as claimed by Westergaard and Resler (1975). The report of the Uthwatt Committee (HMSO, 1942b) recognised that resources should be used in the interest of the country as a whole, but did not accept that this meant land nationalisation or a national physical plan. The Committee did, however, recommend nationalisation of development value which was incorporated into the Town and Country Planning Act, 1947, though this was abandoned by the newly elected Conservative government in 1951.

Detailed control of proposed development is basically *ad hoc* and is carried out by vetting applications for planning permission submitted to the appropriate local planning authority. This is usually done by an applications subcommittee of the authority's planning committee. Permission may be granted with or without conditions or may be refused. An applicant can appeal to the Minister against a refusal or what he considers to be onerous conditions. The Minister, or in certain circumstances his Inspector, decides the appeal after a public inquiry, and may uphold or vary the local planning authority's decision. In dealing with applications for planning permission a local planning authority must have regard to the provisions of their development plan, though they are not confined solely to it. Indeed it may well be out of date. In principle, control is to be exercised so as to be consistent with the plan's intentions. In exceptional circumstances, the Minister may require an application to be referred to him for decision in the first instance, though in practice this power is rarely used. Development which is in breach of planning control is covered by various classes of enforcement procedure, fines and so on (see, for example, Telling, 1973, Ch. 9).

Objections to development plans, compulsory purchase orders, New Town designation, or in connection with an appeal against a planning decision are very important. The value of land depends preeminently on its use. Once granted, the planning permission usually runs with the land, and thus there is every incentive for the developer to obtain the most advantageous planning permission possible. This has led to an increasing number of appeals against refusal of planning permission, or against conditions. Town planning inquiries are quasi-judicial in character, establishing facts by cross-examination of expert witnesses. The proceedings may take several days or in some cases weeks, and several months more will elapse before the decision is announced. The appellant, the local planning authority, any third parties who have made representations, and persons required to attend by the Minister all attend as of right. The Inspector may, at his discretion, allow other people, for example representatives of local organisations, to appear. A public inquiry is 'public' only in the sense that members of the public may witness the proceedings. It is usually poorly advertised, lengthy and confined to weekday office hours. Members of the public are not given the documents and statements of evidence that make up the cases of the conflicting parties. The formal language and quasi-judicial presentation, often with barristers representing the main parties, all serve to make such proceedings cumbersome, inaccessible and difficult to follow. Indeed Levin (1969) went so far as to call it a farce.[22]

In principle the definition of 'development' is a very wide one,[23] but, as already mentioned, there are serious limitations to its control. These are of two kinds: constraints imposed by the general economic structure, and those which pertain to development control procedures. Limitations of this second type are points of detail, but none the less important, and may have serious effects in specific localities. They are listed below.

*Limited Definition of Grounds for Refusing Planning Permission*

An application for development can only be refused on physical planning grounds, and these are narrowly defined. For example, it might be argued that the proposed use would generate too much traffic, or would constitute 'over-development' of the site, or be an inappropriate use of it. It is not possible to refuse permission for luxury flats on the grounds that they are too expensive for the local community, or a private hospital on the grounds that it is a *private* hospital. It may be argued that it is usually possible to 'trump up' physical planning grounds if necessary in order to turn down applications that are unacceptable politically. To some extent this is true, but the reason given for the refusal must bear investigation at a public inquiry should the applicant appeal against the decision.

### *Limited Scope of Conditions that may be Attached to Planning Permission*

Conditions attached to planning permissions must be 'fit and proper' from a planning point of view, again narrowly defined. They might include providing adequate road access to the site, putting in drains, or providing a certain amount of car-parking space. In 1971 the scope of planning conditions was widened somewhat, such that permission can now be conditional upon provision of some building, land or service to constitute a benefit to the local planning authority, over and above the development contained in the proposal. However, it is unlawful to attach planning conditions that require certain kinds of people to be housed in proposed residential accommodation or to be employed in industrial premises, for example.

### *Limited Rights of Objection*

Only the applicant can object if planning permission is refused, or 'onerous' conditions attached: Others may not make objections to such decisions and it is not possible for anyone to object to permission being granted.

### *Enforcement Procedure*

In certain circumstances the enforcement procedure is not an effective deterrent against breaches of planning control. There are fines imposed for unauthorised development but these may be very small sums compared to the advantages to be gained, as, for example, with unauthorised demolition which may result in increased potential site coverage or speed and ease of redevelopment.

### *Individual Consideration of Isolated Applications*

Each application is to be judged individually, on its merits for the particular site, and to some extent in isolation from other applications for the surrounding area. The cumulative impact of several applications to convert houses into flats, or into hotels, for example, tends not to be considered. If previous applications for the same change of use have been successful, consistency requires that further applications be allowed also. In principle it would be possible to refuse further applications on the grounds that 'sufficient' cases of the change of use had already been allowed, though in practice it may be difficult to define 'sufficient'.

## Changes of Use that do Not Require Planning Permission

Some changes of use do not require planning permission where the change is not considered substantial and where the uses fall into the same Use Class, as, for example, a change of use from a grocer's shop to a boutique or antique shop, from a hostel into an hotel, or from rented to owner-occupied housing.[24] Some of the Use Classes are very wide, which means that whole areas can change their character in a fundamental way, yet this change is beyond the scope of planning control. This especially applies to 'gentrification', where low- and medium-cost housing is improved and let or sold at very much higher prices, so that the original tenants can no longer afford it. This process is accompanied by changes in local shops. A plumber's merchants, second-hand furniture shop and snack bar may thus become an antique shop, wholefood grocery and wine bar.

## Reactive Nature of Development Control

By and large a local planning authority can only react to applications submitted, using development control as a defensive weapon. Development depends upon somebody being willing and able to develop a particular site for a use acceptable to the planning authority. If there are no suitable applications forthcoming, the site will remain undeveloped.

This discussion of the detailed limitations of development control procedure leads on to more general and much more fundamental constraints on the British system of land-use planning and control of development in the context of a capitalist economy. The key to the problem of securing control over development is the private owner-ship of land, and private initiative in development. As stated earlier, commercial development is a business venture attracting capital to the extent that it compares favourably with other forms of invest-ment. Planning legislation cannot make private development happen if there is no interest on the part of developers and financiers in a particular type of development, or in a particular location.

The market in land, and its high price, especially in London, limits the ability of local authorities to carry out development themselves. 'Land for planning permission' was a well known formula used by the (then) London County Council to solve its poverty, according to Marriott (1967). A classic example was the proposed St Giles Circus roundabout. The necessary land was fragmented and privately owned by several landowners, and the council could not pay the high price that was asked for it. Hyams offered to assemble the site so that the London County Council could go ahead with their road proposals, and in return he was to get planning permission to build Centre Point.[25] This policy had the advantage of being mutually satisfactory to both parties, and avoided breaking the law against the sale of planning permission. It also resulted in far higher densities than the London Development Plan allowed, for Centre Point was built at a plot ratio[26] of 10:1 as compared to the 'permitted maximum' of 5:1.

The private ownership of land is at the root of two problems identified in the White Paper *Land* (HMSO, 1974), the basis of the Community Land Act, 1975. First, the system of planning control is almost entirely negative, and second, there is the problem of compensation and betterment. Indeed, this was referred to as long ago as 1909 by Lloyd George:

> The growth in value, more especially of urban sites,... due to no expenditure of capital or thought on the part of the ground owner, but entirely owing to the energy and enterprise of the Community...is undoubtedly one of the evils of our present system of land tenure that instead of reaping the benefit of the common endeavour of citizens a community has always to pay a heavy penalty to its ground landlords for putting up the value of their land (quoted in HMSO, 1974, p. 1).

Early attempts to introduce a system of town and country planning, from 1909 to 1932, were frustrated by the problems of compensation and betterment. Local authorities were required to pay compensation if they prevented the development of land, but provisions designed to enable these authorities to receive a share of 'betterment' when development went ahead were largely ineffective. In 1942 the Expert Committee on Compensation and Betterment was appointed 'to make an objective analysis of the subject of the payment of compensation and recovery of betterment in respect of public control of the use of land' (HMSO, 1942b, para. 1). The Committee assumed that national planning 'directed to ensuring that the best use is made of land' involves 'the subordination to the public good the personal interests and wishes of landowners' and that a purely individualistic approach to land ownership 'operates to prevent the proper and effective utilisation of our limited national resources' (para. 17). Confiscation was rejected as not providing a solution to the compensation and betterment problem, 'but only an expression of a particular political theory' (para. 39). It was argued that political controversy would involve lengthy debate; that 'financial operations which might be entirely out of the question' could be involved (i.e. compensation); that complicated administrative machinery would be necessary (para. 47). However, the Committee recommended 'the acquisition by the State of the development rights in undeveloped land' (para. 48).

The Town and Country Planning Act, 1947, drew substantially on the work of the Uthwatt Committee, and represented the first comprehensive approach to the related problems of land-use planning, compensation and betterment. The main provisions of the Act were that all development required planning permission, with no compensation payable if permission was refused. A £300 million fund was established, from which compensation would be paid to landowners whose claims for deprivation of the 1947 development value of their land were accepted. A development charge was payable before development started, set so as to equal the difference between the value of the land in its existing use and after develop-ment. As a result of this nationalisation of development value, all land was supposed to change hands at existing use value, which meant that local authorities could buy land relatively cheaply for their own purposes. The system was criticised as removing the incen-tive for owners to develop their land, because owners who were going to have to pay a development charge should, in theory, not have paid more than existing use value in the first place. In practice, land changed hands at more than existing use value, though at less than full market value, with the expectation of development. It is suggested that 'it is likely that withholding land from the market may have arisen from the expectation that the system would be aban-doned if there were a change of government' (HMSO, 1974, p. 2). Nevertheless, there was a substantial property boom after the Second World War (Franks, 1974) despite restrictions on building materials and the belief in some quarters that the provisions of the 1947 Act would completely stifle development (Royal Institute of

Chartered Surveyors, 1947).

The betterment provisions were in fact repealed following the change of government in 1951, though the Town and Country Planning Acts of 1953 and 1954 did not alter the provisions under which public authorities acquired land at a price largely excluding development value. Pressure against this two-price system led to its abolition in 1959. It was precisely this denationalisation of development values in the 1950s which led to a rise in land values, as a consequence of which 'horse trading' with developers was forced on local authorities, as already mentioned in the case of Centre Point.[27] Many more examples are to be found in the 'planning bargains' of the 1970s.

### The Scope of Planning Bargaining

Land values partly depend on the position and the attributes of the sites in question, and partly on what they are and can be used for. This latter is decided by the state by means of planning zoning, which is thus a crucial factor in setting land values and prices. Under economic conditions favourable to developers, where land prices are rising, there are particularly large profits to be made from commercial redevelopment, and planning authorities may be able to extract some additional 'community benefit' in return for planning permissions, for in these circumstances the local planning authority has the upper hand. In granting planning permissions for offices, for example, it is literally giving away hundreds of thousands of pounds at a stroke of the pen. Ambrose and Colenutt (1975, pp. 82-3, 100-2), Counter Information Services (1973) and Wates (1976) all quote examples of proposed office development in inner London where this would have, and in some cases has, happened.

Jowell (1977) reports the results of a survey of 87 local authorities with powers relating to development control, where 44 authorities admitted to achieving planning gain on applications for commercial development, with a considerably higher proportion (over two-thirds) in London, especially the inner boroughs (11 out of 12). Planning gain was defined as 'the achievement of a benefit to the community that was not part of the original application (and was therefore negotiated) and that was not of itself normally commercially advantageous to the developer' (p. 418). The definitions of 'benefit' and 'community' are crucial here, and not necessarily straightforward. A variety of planning gains have been achieved: specification of use of property, for example for residential or industrial purposes, dedication of land to public use, as with roads and footpaths, 'extinguishing' existing users, for example non-conforming uses, provision of community buildings, such as a sports centre or community centre, rehabilitation of property, as with non-listed historic buildings, or restoring land after dumping waste, and provision of infrastructure, such as sewers or drains.

A planning gain policy has often been supported by the press and local organisations, both criticising local planning authorities for taking too soft a line with developers, and accommodating their proposals too readily.[28] An editorial in the professional planning journal shows concern for professional integrity in the context of such bargaining. It accepts that a planning gain is highly desirable, provided that:

> the means of obtaining it are not an abuse of privilege... A developer may voluntarily accept such an agreement as the price he has to pay to do what he wants but it is not uncommon for the developer to reluctantly have to accept doing something extramurally so that he may be permitted to do what he has proposed (*Journal of the Town Planning Institute*, 1972, p. 340).

The editorial goes on to warn that due to high land prices development is often appraised 'on critical financial margins', and a developer may abandon a project if required to make a contribution to a planning gain. The conclusion is that a 'wise planning authority will know exactly where to draw the line between effecting a planning gain on reasonable terms as opposed to others too onerous to be reasonable'.

If the commercial development process is to be 'controlled' even in a limited way by securing planning gains for the community — 'loss-making elements' to the developer — this sort of squeamishness may seem out of place. Development control officers should be as adept at calculating financial margins and likely profits to be made if planning permission is granted as the developer's own staff. Development control work tends to have low status within the profession and is considered to be a matter of routine form-filling, less creative than 'forward planning'. Yet it is crucially important in terms of its effects on livelihoods and environment. Harrison (1972) describes the relative isolation of development control sections within local authority planning departments, despite the fact that they take a substantial proportion of staff. He points to the contraditions of the job, and its unattractiveness to newly qualified planners:

> The influx of graduates into planning in recent years would appear to have had little effect on development control thinking, and in fact officers report that development control is an unpopular task among new recruits. One reason why ambitious planners are wary of this type of work may be its dependence upon traditional approaches, and the strict limits which the legally defined task imposes. Younger planners would perhaps rather work on more academic tasks where they can be insulated from the real political processes and can also retain illusions about the social and economic benefits of planning. The limits encountered in implementation can be ignored and there is no need to face up to the real image of planning with the public (p. 269).

Nearly thirty years ago the Schuster Committee report ( HMSO, 1950) on the training and role of the planning profession stressed the importance of administrative skills, but despite this, planning education has continued to underemphasise the financial aspects of development, and administrative or political aspects of planning practice and project implementation.

If the development control process is in any way to fulfil its policing function, it needs a parallel organisation to that of the developer in terms of expertise and resources. But of course the critique goes much further than this. It is the fundamental limitations of land-use planning in a capitalist economy, where the rules of the market make the running, which give rise to this sort of bargaining in the first place. Such bargaining is contrary to the ideal which is supposed to govern planning: namely the pursuit of the 'public interest' without concessions to limited private interests *en route*. The planning gain agreements made possible by the property boom of the late 1960s and early 1970s are clearly *ad hoc* attempts to cream off some of the very large profits for the benefit of the community, only possible in boom conditions. Ratcliffe (1974) offers several criticisms of the approach: local planning authorities may seek to impose illegal or invalid conditions on planning permissions. Unreasonable haggling over proposals may delay development even more than usual. The incorporation of residential units and public facilities as part of development projects may be somewhat arbitrary and conflict with comprehensive planning. Loughlin

(1978), however, reports a case in the London borough of Harrow which he sees to be very successful as far as negotiating planning gain is concerned. But if land, or development value were public property, then all gains except developers' 'normal profit' would be public. As already mentioned, between 1947 and the 1950s increases in land value were *de facto* first almost wholly, and then still in part, in public ownership.

Final plans or development proposals are undoubtedly the product of a process of horse-trading between local authority and private developers and/or owners. This process may not be easy to trace — partly because so much of it goes on behind closed doors, though there may be some leaks — and not least because its likely outcome will probably have been partly anticipated by the local authority from the outset, with the local authority councillors and planners 'taking account of realities' even before they first set pen to paper. Hence the 'corruption' of the planning process by the brute facts of ownership, land values and market pressures may be partly invisible, and may only be seen by comparing what is proposed with what might have been, or could have been proposed were it not for those brute facts.

Public intervention in the economy in general, and in land-use planning in particular, has been steadily increasing, especially since the Second World War. The general introductory remarks provided in this chapter will be amplified in later chapters, for three areas stand out as requiring more detailed treatment: the changes which have been taking place in local government, the organisational context of urban planning, the role of professional planners as state employees, and the various possibilities this holds for different individuals, and the scope for public influence over planning objectives and details of implementation. Before going into these substantive areas in more detail, however, I want to consider the usefulness of the academic literature, and to outline theoretical perspectives which have a bearing on these issues. This theoretical work will be the concern of the next chapter.

**Notes**

1. See, for example, Asworth, 1954; Bell and Bell, 1969; Cullingworth, 1976; and Glass, 1959, pp. 393-401.
2. The Housing, Town Planning etc. Acts, 1909 and 1919, permitted the planning of land outside urban areas needed for development; the Town and Country

Planning Act, 1932, permitted the planning of urban areas; the Town and Country Planning Act, 1947, made provision for town planning in the whole of England and Wales. Other associated legislation available to local authorities in connection with land-use planning proposals includes the Highway Acts, Housing Acts, Public Health Acts, Civic Amenities Act, 1967, Community Land Act, 1975, and so on.

3. For a résumé of this activity see, for example, Cullingworth, 1976; Hall, 1973; Marriott, 1967; Osborn and Whittick, 1969; and Palmer, 1972.

4. The local planning authorities are county councils, county district councils, the Greater London Council and the London boroughs. See Telling (1973, pp. 47-8) for an account of how the various functions are divided between the different authorities. Some very limited forms of development are exempted from requiring planning permission.

5. The terms 'middle class' and 'working class' have given rise to much comment and argument concerning their definition and continued usefulness, as discussed for example by Westergaard (1972, pp. 144-8). For present purposes I am using 'working class' as a convenient shorthand, broadly to mean people in skilled and unskilled manual work, and routine clerical work, and 'middle class' to mean professionals, managers and executives. For discussions of class see Allen, 1977; Giddens, 1973; Hunt, 1977; and Wright, 1978, Ch. 2, for instance.

6. Hounslow Hospital Occupation Committee, Elizabeth Garrett Anderson Joint Shop Stewards' Committee, Plaistow Maternity Action Committee, Save St. Nick's Hospital campaign are examples from London.

7. There is a considerable debate as to the nature of the contradictions in the capital accumulation process which push the capitalist system towards economic crisis. Four main explanations concern the tendency of the rate of profit to fall (Yaffe, 1973), the problem of underconsumption (Baran and Sweezy, 1966), a falling rate of exploitation resulting from rises in wages (Glyn and Sutcliffe, 1971) and the contradictory role of the state in accumulation (O'Connor, 1973; Yaffe, 1973). Wright (1978, Ch. 3) reviews these various formulations, as do Holloway and Picciotto (1978, Ch. 1).

8. Details of measures adopted in Britain in an attempt to manage the economy are given by Broadbent (1977, Ch. 2), for example.

9. These include the Hunt Report (HMSO, 1969a) which led to the designation of development areas and intermediate 'grey' areas; the development of New Towns for their allegedly stimulating effect on the local economy, for example Peterlee, Telford, Leyland-Chorley; beneficial rate levels for relocating firms; the provision of industrial estates; the setting up of the Location of Offices Board, and so on. See also Harris, 1966 and Holland, 1976.

10. Disparities in levels of amenity and services are not necessarily noted locally or, if noted, may not be considered relevant or important, as argued by Mellor (1975, p. 289).

11. According to the *Department of Employment Gazette* (May 1978), workers involved in temporary employment subsidies were estimated at 408,151 for the period August 1975 to March 1978 (p. 544). The total number of jobs created or subsidised by the Job Creation programme and Community Industry and Small Firms Employment Subsidy schemes was estimated at about 150,000 (p. 544). Further, a Regional Selective Assistance programme was expected to create nearly 275,000 jobs and safeguard another 115,000 during 1978/9.

12. For accounts of local authority housing see, for example, Community Development Project (1976) and Conference of Socialist Economists (1975). Access to local authority tenancies is based on an assessment of 'housing need' based on people's current circumstances — shared facilities, overcrowding, poor physical condition of accommodation, and any special needs concerning health or disability. Some priority is also given to residents awaiting rehousing from proposed slum clearance areas, and to 'key workers'. Who counts as a key worker will vary from

place to place, but includes public service workers and employees of private firms which would not locate in the area or have to move out of the locality if local authority housing were not available for employees. Council housing waiting lists are often taken as an indication of housing stress for practical purposes, though they provide a conservative estimate of demand. They do not include people who are ineligible for council housing, such as newcomers to an area who do not have the required period of residence — varying between 1 and 5 years, depending on the authority. Many people do not bother to register, deterred by the length of time they expect to have to wait before being made an offer. Moreover the waiting lists give no indication of the adequacy of housing or the standard of facilities available. Local authorities are obliged to house people with children, and the 'homeless', and in practice this may amount to no more than temporary 'bed and breakfast' accommodation in a hostel or hotel. Mortgages for owner-occupied housing are worked out for a period of 25 years as a rule. They are not readily available to people over 40 years of age, or for older houses or converted flats. One hundred per cent mortgages are uncommon, so prospective buyers need savings and a secure income which is not likely to decrease in the future. Sums loaned are generally computed at twice a man's salary, plus a wife's salary. At current average income levels and house prices, this would not buy a one-bedroomed flat in inner London.

13. 'Gentrification' refers to the improvement of property, which will then command a higher price than before, and will only be accessible to people in upper-income brackets. The term was coined by Glass (1970). See also Conference of Socialist Economists (1976, pp. 44-9), and Counter Information Services (1973, pp. 40-4, 62-4). Raban (1975, Ch. 8) gives a nice description of the gentrification process.

14. Some goals that have been written into land-use planning documents are not capable of achievement under the present system. For example, *The South East: a Framework for Regional Planning* had the goal of providing for 'personal satisfaction in work'. A goal of the *Greater Peterborough Master Plan* was 'to offer the highest possible quality of social and family life for all' and a goal of the *Teeside Survey and Plan* was that of 'establishing satisfactory economic conditions'.

15. Very little work appears to be published on the subject of non-professionals reading maps and plans. See Stringer and Taylor, 1972.

16. F.J.C. Amos, the then Chief Planner for Liverpool, and President of the Town Planning Institute from 1971-2.

17. The total annual net investment of UK insurance companies was £972 millions in 1970 and £1,618 millions in 1972. During this period the companies invested about £200 million per year in property (12-13 per cent), as reported by Counter Information Services (1974, pp. 8, 9). Examples of take-overs are Sterling Estates, taken over by Royal Insurance, and Edger Investments by the Prudential.

18. Total investment exceeds £700 million per year, 15-20 per cent on property (Counter Information Services, 1974, p. 9).

19. Office development permits (ODPs) were introduced in 1965 in an attempt to impose central government control on office development in London. There were speculative ODPs, and those linked to a specific prospective tenant, whereby the development company had to justify the need for office space. Planning permission could be granted by the local planning authority without a valid ODP, but actual office development could not go ahead without one. ODPs were required for office development in excess of 10,000 square feet originally, raised to 15,000 square feet in May 1977, and subsequently 30,000 square feet. Pressure from the development lobby has caused the Conservative government to abolish the ODP system altogether.

20. There are two exceptions to this general statement: compulsory purchase of land for housing by local authorities, which must be approved by the Minister and compensation paid at 'market price', which will undoubtedly reflect an element of blight; the provisions of the Community Land Act, 1975, which offered some con-

cession to local authorities, though soon to be scrapped by the Conservatives.

21. There have been Master Plans, Development Plans, and now Structure plans and Local plans, and in London also Borough Plans, under the various Town and Country Planning Acts. For their scope see Heap, 1973.

22. Nevertheless, well organised pressure groups have managed to disrupt public inquiries to draw attention to their points of view. In the case of the inquiry into the proposed alterations to Archway Road in London (September 1976-October 1977) local residents successfully demanded that the inquiry be held in the evening, and in a local hall, making it a little more accessible to the public. Since then this has happened in other inquiries. Vociferous opposition to various proposals for the redevelopment of the Coin Street site near Waterloo has led to one of the interested development companies — Heron Corporation — withdrawing from the inquiry, arguing that 'enquiries of this nature are wholly unsatisfactory vehicles for processing development proposals of national significance' (S. Haywood, 'The Battle for Waterloo', *Time Out*, 22-28 June 1979, pp. 7-8).

23. Defined as '(a) the carrying out of building operations, engineering operations, mining operations or other operations in, on, over, or under land, or (b) the making of any material change in the use of any buildings or other land' (Town and Country Planning Act, 1971, ss.290(1) and 22(1)).

24. The Town and Country Planning (Use Classes) Order, 1972, specifies 18 Use Classes. Article 2(2) of the order concerns definitions: the word 'shop', for example, means a building used for the carrying on of any retail trade or retail business wherein the primary purpose is the selling of goods by retail.

25. The proposed roundabout was never built. Other notable examples in central London are the Stag brewery development opposite Victoria Station, and Euston centre. See Counter Information Services, 1973; Elkin, 1974; Marriott, 1967, pp. 121-44, 181-95; and Wates, 1976.

26. Plot ratio refers to the relationship between the amount of permitted floor space and the area of the site. A plot ratio of 1:1 would apply to a single-storey building covering the whole site, a two-storey building covering half the site, a three-storey building covering a third of the site, and so on.

27. More recent attempts to deal with these problems include the Land Commission set up by the 1966 Labour government which subjected development gains on land to a flat rate tax of 40 per cent. It was repealed on a change of government in 1970, before the Land Commission had been given wider powers of acquisition which the Land Commission Act, 1967, provided for after a 'second appointed day'. Then there is the development gains tax introduced in the Finance Act, 1974, which made provisions for varying rates of income tax or corporation tax on development gains. For details see Johnson, 1974. The Community Land Act, 1975, is a more recent measure, and the Development Land Tax Act, 1976, which makes provision for a tax on all development profit — 66.6 per cent over £10,000 and 80 per cent over £150,000 (Massey and Catalano, 1978, Ch. 8). However, very soon after its election victory in June 1979, the present government announced its intention to repeal the Community Land Act and to introduce lower rates of development land tax.

28. For example, C. Booker and B. Gray, letter, 'The Profits from Development', *The Times*, 12 Dec. 1973; 'Giving the People a Slice of the Property Cake', *Observer*, 2 Dec. 1973; D. Brewerton, 'Glad Tidings for the Southwark Developers', *Daily Telegraph*, 5 Feb. 1974; P. Riddell, 'Public Interest and Private Gain at Hay's Wharf', *Financial Times*, 26 Nov. 1973.

# 2 THEORETICAL APPROACHES TO URBAN PLANNING

This chapter deals with a consideration of four theoretical perspectives which purport to provide some explanation of aspects of the land-use planning system. The discussion is amplified in the following three chapters, in connection with specific themes — the organisational context of planning, the role of the professional planner, and the scope for public involvement. Here it is my purpose to outline the main points of the different approaches: pluralist, bureaucratic, reformist and Marxist. Before doing this, however, a few general remarks are in order.

All four theoretical approaches to be considered here are concerned directly or indirectly with the distribution of power in society, and the mechanisms for effecting possible changes in existing power relationships. Each perspective focuses on the people, institutions or structures which are assumed to constitute the locus of power. Thus, in the case of the pluralist approach, the initial assumption is that in liberal democratic societies power lies with 'the people', who organise themselves to present their views, ideas and protests to government, which responds to the pressure brought to bear on it in this way. Theorists who stress the power of public bureaucracies start with this assumption, separating out the roles of politicians — Members of Parliament and local council members — and administrators, civil servants and local government officers. Some of these writers emphasise the influence of bureaucrats, as compared to elected politicians. On this view, a challenge to the power of administrators can come about through democratic political activity. What I have called a reformist view attributes power to government, and beyond it, to business interests. Changes in the existing distribution of power will come about by political activity on the part of disadvantaged groups, and ultimately national government's willingness to make changes. These are marginal changes, however, limited by the overall distribution of power. Marxist theories focus on the distinction between owners of the means of production, and the propertyless, who are forced to sell their labour power in order to survive. Power is held by property-owners, though non-owners may be disruptive, and challenge their exploitation, either by threats or action. The state has a close relationship to the owners of the means of production, though it does not just crudely reflect their interests. It may need to make concessions to workers, in order to stave off strikes, for example, and to secure continued co-operation. Workers' capacity for conflict is limited by their need to earn wages. Thus the four perspectives have an entirely different focus of interest, though they are all concerned with the distribution of power.[1]

These four headings are used for the sake of convenience. Such labelling may give the impression that they apply to distinct, clearly articulated bodies of theory, though in fact this is not the case in any strict sense. The purpose of the headings is merely to impose some structure within which different contributions can be discussed. It should be pointed out that these theoretical perspectives are not directly comparable. They have been employed for different purposes and have not always been formulated specifically in the context of urbanism, though applications to this context can be developed. Between the various approaches there are areas of overlap and areas of divergence, such that there are both competing and complementary aspects. The different perspectives may be distinguished in terms of a number of criteria: the extent to which they take account of the wide range of factors involved in urban situations and land-use planning, and the degree to which the interplay of spatial, social, political and economic aspects is acknowledged and taken into account. Associated with this is whether they recognise separate sub-disciplines within the social sciences — economics, politics, sociology, social administration and so on — and the way in which this recognition delineates the field of 'the relevant'. Another difference concerns implied commitment to political action, and given this, the kind of political activity in question — using 'political' in the broadest sense, not only to mean party politics.

There is an irregularity of treatment and terminology in these discussions which it is important to acknowledge, arising out of variations in conceptualisation and language on the part of the different writers whose work I am considering. These variations are reflected in my commentary, which attempts to use and follow the terminology of the theorists themselves. Though I have tried to avoid gross inconsistencies, this unevenness of treatment should be pointed out, and borne in mind.

### The Pluralist Approach

This perspective has been particularly pervasive and influential, and merits consideration in some detail. To some extent several of the ideas associated with it have passed into conventional wisdom in what Dennis (1972) calls 'the modern theory of democracy', popularised by Schumpeter (1943) and subsequently such writers as Lipset (1960), Dahl (1961, 1967) and Polsby (1963), with its stress on the passivity, even apathy, of the individual, whose political activity is generally restricted to voting in elections. The overall aim of the political system is seen as the maintenance of the stability of society as a whole, on this view.

Although there are several slight variations according to different pluralist writers, fundamental common themes can be distinguished, and this perspective is also associated with the work of Bell (1960) and Galbraith (1957, 1967), and studies of local power structures focusing attention on different group influences over decisions concerning 'key issues' undertaken by Dahl and Polsby, and Banfield (1961). This approach has been developed with reference to the American political situation, but in Britain similar assumptions have been expressed or implied (Crosland, 1956, 1962; Finer, 1955, 1956; McKenzie, 1958, 1964), and applications made to industrial relations (Fox, 1973; Selekman and Selekman, 1956).

On the pluralist view, society consists of a collection of interest groups competing for control over government action through the electoral process. Pluralists 'see society as fractured into a congeries of hundreds of small special interest groups, with incompletely overlapping memberships, widely differing power bases, and a multitude of techniques for exercising influence on decisions salient to them' (Polsby, 1963, p. 118). Power is said to be distributed in a diffuse way so as to guarantee that no one group can dominate any particular segment of society. Different groupings of individuals align themselves in various combinations according to the issues. If a particular interest threatens to gain the upper hand, opposition groups will emerge to challenge the powerful group, and thus the equilibrium will be maintained. Playford (1971) distinguishes two slightly different roles for government. In the 'balance of power' variant, government is forced to accommodate itself to a number of conflicting interests, among which a rough balance is maintained. This is differentiated from the 'referee theory' of pluralism where government supervises and regulates the competition of interests so that none will abuse their power to gain mastery of some section of social life. The political process is characterised as 'open' and democratic, with easy access to decision-makers and numerous channels of communication for individuals and groups. In practice only a minority may feel strongly enough about a particular issue to be politically active, but in theory there is nothing to stop anyone who wants to do so from becoming active. All it needs is a little effort.

There are three central themes associated with this perspective, as summarised by Dunleavy (1977b). First, the polity is viewed as 'a weak unit lacking in any developed ideology or particularly separate identity, operating in an environment of strong external influences and controlled by politicians who concentrate overwhelmingly on building and maintaining an electoral majority' (p. 194). Such a polity, it is argued, will be very responsive to pressures from the community. Second, there is the assumption that influence is unidirectional: that public opinion plays a decisive role in securing changes in public policy, though the possibility of indirect influence being exerted by elite groups on elected politicians, for example, is rejected. Third, political activity is determined by people's interests. Since it is those people with an interest in the outcome of a political decision who become politically active, and since the weak polity is responsive to external pressures, the outcome will faithfully reflect the balance of interests in a community. Broadbent (1977, p. 205) sees pluralism as a market theory, and claims that it is 'the dominant accepted social theory in the UK'.

> It parallels the neoclassical perfect market ideas of 'fairness' (in competition) 'balance' (between different groups or classes) and 'diversity' (i.e. many entities having legitimate claims on the market).

> The theory...imposes on social policy (the state) the task of ensuring that no single group ends up in an excessively privileged (monopoly?) position (Broadbent, 1977, p. 205).

In support of this view of the distribution of power in contemporary Britain, a number of points can be made. First, there has been a remarkable increase in pressure-group activity in recent years, roughly since the mid-1960s, at both national and local levels, and this is nowhere more noticeable than in the sphere of land-use planning. There has been opposition to the proposed location of motorways, international airports, large-scale power stations and tanker terminals; campaigns for safer roads, more playing space, additional housing, rehabilitation of existing housing rather than demolition and redevelopment, and so on. Much of this local pressure group activity has had some measure of success, and several guides to pressure-group organisation and accounts of campaigns have been published (Crosby, 1973; Davies, 1972; Dennis, 1972; Ferris, 1972; Hall, 1974; Jay, 1972; Jerman, 1971; Kimber and Richardson, 1974; Perman, 1973; Shipley, 1976; Tyme, 1978; the journal *Community Action*, first published in 1972). Further, two major interest groups in Britain, the umbrella organisations of the Confederation of British Industry and Trades Union Congress appear to be on a more equal footing than ever before. The trade unions are said to be growing in power, increasingly consulted by governments on matters of industrial and economic policy, and willing to back up their claims for better pay and working conditions with threats of strike action, which they are prepared to carry out if necessary.[2]

Then there has been an increase in personal and geographical mobility. With the spread of suburban development, the job mobility associated with an expanding economy of the late 1950s and early 1960s, and the more general availability of personal transport with increasing car ownership, it has been argued that communities based on particular locations are tending to be replaced by communities of interest where members are not joined by their common area of residence, place of employment or social class, but come together due to shared interests in particular issues or activities. Such communities are characterised by their relative fluidity and changing membership, and lack of strong links with a specific location. Webber (1968) calls such a community of interest the 'non-place urban realm'. A similar thread is to be found in one of Pahl's essays (1970, Ch. 7). This kind of community of interest may be taken as evidence in support of the pluralist formulation, with its stress on changing alignments and interest groups, for it involves the forming and reforming of social groups and organisations based on the shared interests of a relatively mobile population.[3]

The pluralist perspective also has an inherent appeal, for it contains an implicit egalitarianism, with its stress on the right and ability of people to organise around issues that concern them. If individuals are thought to have equal status then it can be argued that they have equal opportunities for organising themselves to press their particular interests, and theoretically equal chances that such activity will achieve its objectives. Thus the pluralist perspective legitimises existing social and political arrangements, as fundamentally just and fair, involving the idea that inactivity is a sign of satisfaction with the situation as it stands, or at worst, disinterest. The emphasis is on activity, and it seems probable that it will appeal to people who are themselves active in local or national politics, and who, perhaps as a result of this involvement, tend to underestimate the problems it may present for others, who may be less articulate, or lacking information, expertise and resources.

Despite the attractiveness and apparent plausibility of this pluralist approach, its validity must be seriously questioned, as has been done in general terms by Bottomore (1964) and Lukes (1974), specifically in the American context by Bachrach and Baratz (1962, 1970), Domhoff (1978), Gitlin (1969), Mazziotti (1974), Mills (1956), Parenti (1970) and Wolff (1965), and in the British context by Dearlove (1973), Saunders (1975) and Westergaard and Resler (1975). There are three main criticisms: the question of the equality of the various competing intt*Planning* s, the assumption of the one-way nature of political influence from the bottom upwards, and the stress on activities and associated methodological emphasis on 'key issues' for study.

At a national level pressure groups form to promote particular interests and points of view and to persuade central government to introduce or amend legislation or government practice. There are wide variations among both national and local pressure groups. They differ in the scope and type of issues with which they are concerned; their funds and facilities available, the organisational skill of their members, and their ability to articulate demands and mobilise support. These points emerge from accounts of pressure-group activity at national level by Deakin (1968), Eckstein (1960), Hindell (1965), Newby *et al.* (1978), Potter (1961), and Pym (1973), and at local level by Dearlove (1973), Ferris (1972) and Levin (1971a).

In the realm of land-use planning, for example, there are several national organisations which act as pressure groups on central government including the professional institutes: the Royal Town Planning Institute, the Royal Institute of British Architects, the Royal Institute of Chartered Surveyors, the Institute of Auctioneers and Valuers. The professional bodies are often consulted by govern-

ment and make representations about existing or prospective legislation. They are in the habit of making press statements, submitting
evidence or memoranda giving the official view of a particular institute or association, as for example submissions to the Skeffington
Committee's inquiry into public participation in planning and their
comments on the Committee's report (Royal Institute of British
Architects, 1968, 1970; Royal Town Planning Institute. 1968,
1971b) and opinion concerning the Community Land Act (Royal
Institute of Chartered Surveyors, 1974). Then there is the National
Association of Property Owners, the Country Landowners' Association, the road transport lobby, large contractors and construction
firms, which make up what Colenutt (1975) has called 'the property
lobby'. Colenutt discusses their influence on government policy.
The developer's philosophy is that government intervention in the
property industry inhibits development, holds up economic growth,
and is restrictive and unfair. Of the regulations that cause concern,
four are outstandingly important: rent control, taxation, land
nationalisation and local authority planning restrictions. These
organisations and individuals can be expected to use whatever
opportunities are available then to press their interests. This will
include informal discussions as well as formal lobbying, for these
interests are well represented in both Houses of Parliament, and on
both sides, though more closely connected to the Conservative
Party. Other national groups with an interest in land-use planning
matters include the Town and Country Planning Association, the
Civic Trust, the Council for the Preservation of Rural England, the
National Farmers' Union and environmentalist groups like Friends
of the Earth, Intermediate Technology Development Group, Earth
Resources Research and so on.

Much more numerous in land-use planning are the many local
organisations which are concerned to try to influence local
authorities. Over the past ten years or so, hundreds if not thousands
of groups have formed in different parts of Britain to. attempt to
influence land-use planning policy and practice, and the great
majority of these direct their efforts towards the local planning
authorities. The success of these local organisations appears to be
dependent upon several factors and limited in a number of ways. It is
partly dependent upon the size, organisational ability, expertise and
resources available to the group; partly on the group's aims. Second,
it is dependent on the responsiveness of councillors and planning
officers to the organisation's demands. This appears to be selective,

depending upon a group's aims, expertise and style of communication, for local authorities have their own, usually unstated assumptions as to which interests warrant taking into account. Finally, an organisation's effectiveness is limited by its perceptions of its power, and its assessment of what is worth fighting for, and what is not even worth contesting because there seems to be no chance at all of gaining concessions.

Here it is necessary to distinguish conservation and amenity societies on the one hand, and tenants' and community action groups on the other, and to note differences in personnel, resources, objectives and strategies. Amenity societies are interested in furthering the cause of conservation and preservation of existing buildings, street patterns, open space and so on, the self-appointed caretakers of the local and national heritage and environment. They are likely to have middle-class people as members, people who are articulate, and who have (potentially) useful professional contacts both at work and socially for expert advice and information. They usually organise themselves in a way that is acceptable to local councillors and officers, often with a formal constitution and committee structure, and make their demands through acceptable channels, such as lobbying. Further, there is a formal mechanism for involving representatives of conservationist groups in local planning authority policy-making through Conservation Area Advisory Committees, established under the Civic Amenities Act, 1967. The aims of conservation societies may not always be in accordance with those of the local planning authorities they are trying to influence. The objective of conservation may conflict with business and some farming interests, as for example with the conservation and rehabilitation of old buildings, rather than demolition and redevelopment, and there is also the possibility that conservation societies may conflict with local authorities wishing to collaborate with business over redevelopment proposals. Moreover, some local councillors, particularly those from traditional working-class backgrounds, may oppose conservation proposals as not being in working people's interests.

In contrast to amenity organisations, community action groups are usually formed with the aim of bringing about a redistribution of scarce resources in favour of working-class people. In addition they may have a looser organisational structure, less professional expertise, fewer financial resources, and be forced to resort to more 'militant' tactics of direct action to draw attention to their views.

One additional point concerns political affiliation. Conservation groups often claim to be 'non-political' and may have members from various shades of political opinion. Community action groups are very often explicitly political, and may have links with the Labour Party or some members who are involved in smaller parties on the left — the Communist Party, Workers' Revolutionary Party or Socialist Workers' Party. These differences between organisations are particularly important in the context of a discussion of public participation in planning and will be considered further in Chapter 5. For present purposes it can be noted that the pressure groups involved are not on an equal footing as regards resources, organisational skills and objectives.

A second criticism of the pluralist perspective concerns the stress on activity, and the consequent lack of attention paid to inactivity. Pluralists offer two interpretations of inactivity: it is either attributed to satisfaction with the system as it currently operates, or to disinterested apathy. If no complaints are heard, it is assumed that people are either satisfied with the *status quo*, and have no complaints, or else they are not sufficiently interested to register complaints if they have them. What this approach does not allow for is the fact that people may be inactive because they perceive themselves to be powerless and without influence. There may be dissatisfaction which they do not bother to voice, either because they do not know how to do so or because they feel that government, or whatever organisation would be the likely target, would take no notice. This is described, for example, by Batley (1972) who gives a comparison of participation over housing issues in two areas of Newcastle — Jesmond and Byker, and also by Dunleavy (1977a, 1977b). Sklair (1975) found that previous unsuccessful rent strikes in some areas were an important factor accounting for lack of tenant activity against the Housing Finance Act, 1972, which introduced rent rises for council tenants.

The whole perspective is tilted towards this stress on activities. Studies of community power which use this approach concentrate their attention on what the researchers indentify as key political issues, on the assumption that important issues are visible, that they are the ones which surface in conventional politics, generate controversy, and are ascribed importance by the political system itself, what Lukes (1974) calls a 'one-dimensional view of power'. Dahl (1961, 1967), for example, chose to investigate three issue areas in New Haven: the process of political nomination, urban redevelopment and public education. He identified various interest groups, including businessmen, politicians and professionals, and looked at the views they held regarding these key issues, the tactics they adopted and the arguments they used to press their case, and so forth. Power was conceptualised as an open visible set of relationships and mechanisms that can be observed through noticeable activities, as described by Dunleavy (1977b). There are two related questions to be asked here: whether the concerns chosen by researchers are actually the most important ones, and how and why certain concerns become political issues amenable to this kind of observation, and others do not. Domhoff (1978), for example, argues that Dahl's third issue area, public education, was only marginally of interest to the New Haven decision-makers, because they lived outside the city limits and either sent their children to suburban schools or to private schools. Thus he suggests that examination of this issue would not really test Dahl's contention that New Haven was not governed by an elite.

Crenson (1971) was interested to look at the 'non-issue' of air pollution, which did not surface as a political question for some years, despite its prevalence in urban America. This is in contrast to such issues as poverty and racial discrimination, which clearly had achieved the status of political 'issues'. He compares the experience of two areas: Gary, Indiana, and East Chicago. In Gary, a town dominated by US Steel, industrialists never openly involved themselves in the clean air issue, which emerged in the political arena only very gradually. In East Chicago, by contrast, industrialists took an active interest in this question, and legislation to combat it was passed relatively quickly. Crenson suggests that a pluralist interpretation would attribute more political power to industrialists in East Chicago as compared to those in Gary, since they were more politically active in this issue. His own analysis runs directly counter to this, however. Clean air was slow to be taken up as a political issue in Gary, a one-industry town, exactly because of the economic dominance of US Steel. The firm did nothing to push the issue, as clean air legislation could be expected to increase their operating costs, and the politicians also assumed that the company would have little interest in clean air legislation. Thus Crenson explains a lack of activity on the part of Gary industrialists as a sign of their power, rather than the other way around.

Dunleavy (1977a, 1977b) looked at the issue of high-rise housing, a common form for public housing in Britain in the 1960s, despite its unpopularity with residents and people generally. He chose this question specifically because it did not emerge as an open political issue, and explored the various power relationships in operation, involving local authorities, professional architects, large-scale building and construction firms, and working-class tenants and prospective tenants.

This is what Bachrach and Baratz (1970) call 'the second face of power', and their critique of community power studies rests here. For they argue that power may be exercised by restricting the scope of decision-making to relatively 'safe' issues, creating or reinforcing social and political values and institutional practices that limit the scope of the political process to public consideration of only those questions which are comparatively innocuous. 'Innocuous, that is to privileged interests', adds Miliband (1969, p. 156). Bachrach and Baratz talk of the power of 'non-decision making', of business preventing certain questions from being considered. 'Prevention' carries the suggestion of activity on the part of business, and implies conscious effort to ensure that policies are not questioned or challenged. Westergaard and Resler (1975) take this critique further, stressing that the pluralist approach cannot take account of those issues which do not come into dispute at all. Still more important than the power of 'non-decision making'

> is the power to exclude which involves no manipulation; no activity on or off stage by any individual or group; nothing more tangible than assumptions. They predetermine the range of issues in dispute, and limit it to those in which negotiation, competition or conflict has some practical chance of shifting the balance between the contending parties.

This is power that is 'anonymous, institutional and routine' (p. 247).

Finally, against the pluralist view of the uni-directionality of influence from constituents to politicians and officials, there is the question of the resources which can be deployed by dominant groups against lesser groups. As Dunleavy (1977b, p. 198) puts it, this includes the prevention of the 'accurate perception of their interests by the powerless' and the inhibiting of 'mobilization, organizational development and protest success'. The writings of Bachrach and Baratz and Westergaard and Resler already cited speak to this point, as does Dearlove's (1973) study of local government. Elite theorists like Crenson (1971) and class theorists such as Aaronovitch (1961),

Gutsman (1969), Miliband (1969) and Poulantzas (1973) all point to the power of dominant political groups.

To sum up this discussion, the strength of the pluralist perspective lies in the attention that is paid to the activity of local and national organisations and interest groups — their aims and aspirations, membership and support, activities and tactics, in influencing particular policies. Its several weaknesses, however, outweigh this strength. There is over-emphasis on activity at the expense of any consideration of apparent inactivity. The various organisations do not compete on more or less equal terms. Also there is the erroneous assumption that influence is only one-way, from the mass of ordinary electors to official policy-makers who are constantly anticipating possible loss of popularity and electoral defeat. This approach can be contrasted with a view of the distribution of power which emphasises the positions of politicians and public administrators, and it is to this perspective that I shall turn next.

### The Power of Public Bureaucracy

A stress on the power of public bureaucracy is implicit in studies of local authority decision-making, particularly in the spheres of housing and land-use planning, as in the work of Dennis (1972), Davies (1972) and Rex and Moore (1967). It is also implied in Pahl's earlier work, though it is only fair to note that he has to some extent reconsidered this earlier position more recently (1975, Ch. 13).

Both Dennis and Davies give accounts of local authority decision-making where residents' groups are in conflict with the local councils. The councils in question, Sunderland and Newcastle, are not the weak institutions of some pluralist studies, however, receptive and responsive to pressure from below. In this work it is assumed that local authorities are complex structures, difficult to penetrate and largely impervious to influence from local residents. Moreover, within the local authorities the relationship between councillors and officers does not appear to be in practice what it is conventionally conceived to be in constitutional theory. The councillors themselves had little detailed knowledge about the planning and housing issues in dispute and were reliant on the expertise of the full-time officials of the planning and housing departments. It is not only members of the public who have difficulty obtaining information from the bureaucracy, but also local councillors. The account of the activities of the Millfield Residents' Association given by Dennis, and Rye Hill Residents' Association by Davies, though different in style and scope, are both concerned with the redevelopment of blighted residential areas in Sunderland and Newcastle respectively, and the relationship between residents and officials over the decisions whether, and when, to demolish the houses. From the point of view of the residents, the council's intentions were vague, and appeared to change several times before decisions were finally taken. Information about the future of the two blighted areas was withheld from residents, who lived in an atmosphere of remarkable uncertainty. The planning authorities appear to have been obtuse and insensitive in the extreme to the residents' need for definite information concerning the future of their homes. Millfield Residents' Association repeatedly asked for information from councillors but was referred to the planning officers as the people with all the answers. The council seems to have considered the Residents' Association to be a nuisance, making unreasonable demands for information concerning the council's proposals for clearance, and for explanations of both the policy and the investigation and house survey upon which policy decisions purported to be based. The position of the Rye Hill Residents' Association appears to have been similar.

Dennis sees the demand for public involvement in land-use planning decision-making in the context of administrative justice. Agencies of government must balance the need for efficiency in the provision of services against the expectations of democracy. In principle there are two ways by which dissatisfaction with treatment at the hands of administrators can be expressed: either through local councillors, or MPs at national level, or through the Ombudsman or appropriate tribunal. Dennis underlines the importance of there being some machinery to deal with grievances arising out of decisions made by planning officers, where these are experienced as 'maladministration', to safeguard the interests of people so aggrieved.

Rex and Moore (1967) and Pahl (1975) view the city as a relatively discrete social system and look at the distribution and use of scarce resources, and the constraints operating within it. Rex and Moore focus on the immigrant communities of Sparkbrook in Birmingham, particularly the housing which is available to newcomers. These authors were concerned with the bureaucratic rules of resource allocation, for the immigrant workers did not qualify for mortgages, having no capital, nor for council housing, since they could not meet the five-year residence qualification. Hence they had to rely on the privately rented sector and found themselves in lodging-houses, often in poor condition, with makeshift facilities. The city council was concerned with the deterioration of a neighbourhood which had once been the home of white people, and wanted to stop the spread of lodging-houses in other areas. Ultimately, a private Act of Parliament allowed the council to regulate the condition of houses in multiple occupation, and defaulting landlords were prosecuted. Yet these landlords, many of them of Asian or West Indian origin themselves, were providing a service for immigrant workers, which neither the local authority nor the building societies were prepared to do. Furthermore, Rex and Moore argued, as some immigrants who had been living in Birmingham for the qualifying five-year period became eligible for local authority tenancies, they were discriminated against, perhaps unconsciously, and were offered accommodation from amongst the poorest of the council's housing stock. These authors raised important questions, for they made problematic the actions and interests of bureaucrats in local housing and planning departments, and urged that more housing be made available to immigrants by government, and that non-discriminatory allocation policies be pursued.

This problematic is amplified by Pahl (1975, Ch. 12) who is concerned not only with housing but with the allocation of resources generally. He emphasises the importance of researching into the operations of the managers of the urban system, the local technocrats and 'social gate-keepers' who mediate in the allocative processes and who have the capacity to shape the socio-spatial system. These are identified as housing department officials, estate agents, planners, private landlords, social workers and so on.

This general perspective has several strengths. First, there is acknowledgement of the power and complexity of central and local government bureaucracies, and the gradual expansion and increasing control of government activity in everyday life. Central government expenditure amounted to 50 per cent of GNP in 1973-4, and as much as 55 per cent in 1975, falling slightly in 1976 to about 53 per cent. Roughly 30 per cent of this was disposed of by local authorities, thus greatly enhancing the status of local government officers.[4] In addition, government has played an increasingly interventionist role in the post-war period, taking greater responsibility for providing employment, both directly and through support to firms, negotiating wage levels, investing in infrastructure, underwriting particularly large or risky business ventures, renting large office premises built speculatively by private developers, providing an outlet for the products of some firms and so forth, as mentioned earlier. At a local level there has been major redevelopment of town centres and inner-city housing, suburban expansion, and bypass and motorway schemes, all directly or indirectly the responsibility of local authorities. Hindess (1971) notes the vital importance of the local authority for council tenants. In areas where a high proportion of housing is municipally owned, or where a local authority intends to compulsorily purchase owner-occupied or privately rented housing, people's opportunities are greatly affected by the actions of the managing officials of local government departments.

It is a reflection of this increasing pervasiveness of government activity that there has been such a growth of pressure-group activity at local level. Focusing attention on the power of public bureaucracy points up the difficulty that such groups may experience in their attempts to penetrate these expanding bureaucracies, though local groups are not equally placed in this, as I argued above in connection with pluralism. Some groups are better able to cope with the bureaucracy than others, are less intimidated by the complexity of the organisational structure, have some idea who to see, and what to ask for. This requires knowledge of the way government departments and local authorities operate, the division of labour between departments and the extent of the jurisdiction of any one officer or department. It is useful to know what legislation affects a particular issue, what obligations the local authority has, what it may do at its discretion, and so forth. Then there is the terminology. Discussions with planners, for example, are carried out in a technical language with which they are familiar. To argue a case against them requires information, a knowledge of planning procedure and orthodoxy in order to put a counter-case which they would consider valid.

There are additional points to be made about the apparent plausibility of this bureaucratic approach, regarding the researcher's working environment, which concern ease of access to information and sources of research funds. A researcher interested in a local organisation — especially if this represents some disadvantaged section of the community — will perhaps tend to see the situation under investigation through their eyes, and with their definitions, aspirations and expectations. Hence, if they find the local authority impermeable, or bureaucrats insensitive, this will be reflected in the researcher's conceptualisation. It might be argued that it is precisely this ability to share the perceptions of others which makes for valid research. However, this sympathy may lead to a tendency to attribute considerable authority and control to local government officials, for example, at the neglect of the roles of central government or employers, this attribution flowing from a local organisation's view of local government officials as their chief targets. Moreover, Pahl, following Gouldner (1973), endorses the suggestion that some sociologists who are supported by grants from government departments or national research councils tend 'to combine with...the bottom and top in blaming the middle' (Pahl, 1975, p. 267). He suggests that 'it does seem to be the case from recent British studies that the middle dogs have been the chief target for champions of the underdog' (p. 268). A further related point is the question of access to adequate information. If the emphasis is placed on local organisations at the neglect of how local bureaucracies operate, and how local government policy is formulated, this may be partly because researchers do not always have easy access to local authority decision-making. Then there is the possibility that some academics may be susceptible to a view which stresses the power of public bureaucracies partly as a result of their own personal experience of large university and college bureaucracies.

If one is interested in local allocation and decision-making procedures, then the main focus of attention will be on local authorities. The research interest narrows down the field of the relevant and pinpoints certain issues, giving a secondary status to others which are not of immediate concern, or taken as given. Dennis and Davies, for example, are interested in local housing issues and concentrate on the particular relationship between local organisations and the respective local authorities. This interest restricts their viewpoint, and does not incorporate more general questions concerning the relationship between land-use planning and the land market, for example, or central government policies on housing standards or housing finance. Similarly, Eversley appears to completely overlook social and economic factors when he writes of black housing areas of London thus: 'Notting Hill, Brixton or Willesden are not the creation of reactionary politicians. They are signs of incompetence, or unintended by-products of what were only ten years ago declared to be laudable plans like slum clearance' (1973, p. 219).

Pahl's stress on the importance of investigating the role of the system managers is not inherently so restricted as Dennis's and Davies's approach. It originally includes as 'managers' people in a range of potentially relevant organisations in both the public and private sectors, in a particular locality. These managers were later redefined to include only local government officials, and this more restricted managerialist thesis implicitly asserts the overriding importance of local government in the allocation of resources and facilities. Elliott and McCrone (1975, pp. 31-6) have attempted to get away from what they consider to be undue one-sidedness in their study of the role of private landlords in Edinburgh, while accepting the overall usefulness of the original, more extended definition of managerialism. Similarly, Norman (1975, p. 67) argues that the restricted version of the managerialist thesis of Dennis and Davies focuses too narrowly on the local government officials to be an acceptable proposition about the organisation of resources within urban society as a whole. Mellor (1975, pp. 283-4) makes a similar point concerning the limitations of what she calls the 'urban system approach' of Rex and Pahl, where the concern is with 'domestic issues': poverty and community development in the inner cities, race in twilight areas, and 'the failure of the local executives in their professionalism to redress inequality on the ground'. All these issues must raise further questions about national policies for investment and growth, and the metropolitan society's command over international resources. This is especially so with redevelopment and housing where the 'personal world of home and family life is at the mercy of international swings of fortune in that policies for housing expenditure are governed by national status in world markets'. The most important of 'urban infrastructural investment' is not in this respect 'urban'. This point is echoed by Saunders (1978) criticising Rex and Moore's conceptualisation of 'housing classes' as inadequate for an understanding of the housing problem.

Pahl (1975, p. 265) has reconsidered his original formulation and admits two weaknesses: that the approach 'lacks both practical policy implications and theoretical substance'. A stress on the power and impermeability of public bureaucracies often focuses on their lack of information about, and apparent lack of interest in, the needs of local residents as defined by local people themselves. The tendency is then to argue for better communication between professional administrators and residents. This is sometimes presented as a need for the managers to be more in touch with the needs and aspirations of the people they are managing (Davies, 1972), as a need for more resources (Rex and Moore, 1967, pp. 270-1) or the need for existing resources to be better utilised (Falk and Martinos, 1975). The theoretical weakness stems from a belief in the independent influence of the managers, narrowly defined as local government officials, which implies a separateness of local government officers from the bureaucratic control of central government, from the political control of local councillors, and from control by private capital. Norman (1975, p. 74) criticises Pahl in a more fundamental way. For despite his reservations about the managerialist thesis noted above, Pahl allows that the urban managers 'play a crucial mediating role between the state and the private sector and between central state authority and the local population'. Norman argues that the implication of this statement is that the managers somehow stand outside the state, and are not a part of it themselves. This point will be taken up in more detail below in the context of the political economy of urbanism. For present purposes it is sufficient to note the problems that this conceptualisation raises.

To recap, the strength of this approach lies in the attention it pays to the ways in which government bureaucracies operate, their power, the complexity of their structure, and the fact that they are not necessarily responsive to pressure from below. Because of this it is considerably stronger than the pluralist approach. As has been noted, the scope of Dennis's and Davies's work is restricted by their interest and is limited to the relationship between the local organisations and the respective local council, resulting in an anti-bureaucracy-cum-profession diagnosis, in which the only structure recognised is an economically empty structure of authority. Land-use planning decisions are taken by officers and councillors according to their prejudices, convictions and technical expertise, with little regard to the interests of 'the planned'. On this view, presumably, cities just happen by accretion of such decisions.

Both the pluralistic and bureaucratic approaches treat conflicts of interest as taking place within an economic vacuum, without any reference to the fact that in a capitalist society governments operate and make decisions within the constraints of a capitalist economic system. The pluralist perspective, for example, does not suggest that owners of capital have more power than non-owners. Nor does the approach which stresses the power of public bureaucracy question the relationship between local and central government and private business interests, as, for example, with the redevelopment of inner-city areas. It is as if the economic context of political decision-making is taken as a non-problematic 'given', a neutral background against which political and cultural differences are worked out, an influence which can be taken completely for granted. Basic facts of life in Britain and America, such as substantial inequalities in wealth and power, the fundamental role of private ownership of property and market forces, the power of business interests and assumptions about their validity and continuance, are glossed over or ignored altogether, presumably because they are not seen to be relevant. One of the aims of this book is to demonstrate that such issues are very relevant and that it is essential that they are incorporated into an account of urban processes for that account to be a cogent one.

This leads me to consider a third theoretical perspective, reformism, which does not make this omission, and which does attempt to relate political decision-making, both in land-use planning and more generally, to an analysis of the economic structure of capitalist society.

## Reformism

The third perspective to be discussed here is what for want of a better term I have called a reformist approach. This is a label which should be applied with some caution to the various writings and activities so designated, yet such an approach seems to me to be sufficiently clear for it to be differentiated from the others I am considering, though there is an obvious overlap with Marxist approaches discussed below. It is the perspective of some academics, some activists in the trade union movement, Labour Party and left-wing political groups, and some aspects of social administration — at least the declared objectives, if not always the actual outcomes of social policy. It embraces a very mixed collection of work and writings, is associated particularly with poverty, homelessness and poor environment, and characterised by humanitarian, egalitarian aims, involving positive discrimination in the redistribution of resources.

The strength of a reformist perspective is the recognition given to basic structural inequalities of power, influence, income and wealth in modern Britain in a way that the pluralist approach does not, and indeed perversely ignores. The theoretical underpinnings of this approach are derived from two main sources: a humanistic liberal egalitarianism, and some variant of Marxism, and this dual legacy gives rise to varying degrees of recognition and importance being attached to the problem of reform (Cowley *et al.*, 1977; George and Wilding, 1976, Ch. 4; Marris and Rein, 1967). This perspective entails critiques of persisting inequalities in Britain and America, explanations of how these have come about, and how they are reinforced by prevailing institutions, and thus perpetuated.

Harvey's work (1973), which has been particularly influential, is concerned with the issue of inequality, specifically in the context of urban areas, and mechanisms which govern the distribution of income within the urban system. These include the price of accessibility to facilities and the cost of proximity to sources of noise and pollution, the ability of groups with financial resources and education to adapt to changes in the urban system more rapidly than other groups, and the uneven distribution of externality effects, though in principle these can be both positive and negative. In the past geographers have not taken up normative questions of distribution, and location theory has relied on the criterion of efficiency. Harvey sees his role as an academic as enhancing the understanding of urban processes in order to facilitate a more just distribution of scarce resources.

> we need to move to a new pattern of organisation in which the market is replaced (probably by a decentralised planning process), scarcity and deprivation systematically eliminated wherever possible, and a degrading wage system steadily reduced as an incentive to work, without in any way diminishing the total productive power available to society. To find such a form of organization is a great challenge, but unfortunately the enormous vested interest associated with the patterns of exploitation and privilege built up through the operation of the market mechanism, wields all of its influence to prevent the replacement of the market and even to preclude a reasoned discussion of the possible alternatives to it. (1973, pp. 115-16).

> The social organisation of scarcity and deprivation associated with price fixing markets makes the market mechanism automatically antagonistic to any principle of social justice (p. 116).

This raises the question of the scope for redistributive measures within the land-use planning system: who is intended to benefit from planning schemes, who actually benefits, and whether one person's or group's benefit always involves another's loss. In Britain, according to local government rhetoric, planning schemes will benefit the community as a whole, and this claim is often used as an 'argument' to resist interest-group demands as sectional, and hence secondary to the wider public interest. In fact it tends to be those groups who are advantaged by the workings of the market who have most to gain from land-use planning, which works to reinforce and consolidate their position, for example enhancing the environment of middle-class residential areas by planning measures like traffic management schemes, declaration of Conservation Area status and so on which serve to make such neighbourhoods more exclusive still, and hence more expensive (Glass, 1959; Goodman, 1972; Lamarche, 1976; Lojkine, 1976; Pahl, 1975, p. 148). Such people are much better able to export external diseconomies — the unwanted airport, motorway, through-traffic route, hypermarket — and to 'protect' their area by campaigning and lobbying (Harvey, 1973, Ch. 2; Hillman, 1971; Levin, 1971a, Mishan, 1967, Parts 2 and 3). But it is more than just being articulate, of course. Middle-class people, especially those with some form of unearned or inherited wealth, can afford to live in places they deem desirable, and can consistently outbid others who are less well off. The dramatic rise in property values in places like Barnsbury in London (Ferris, 1972) after its 'discovery' and subsequent gentrification by middle-class professionals is a case in point here.

Within the constraints of the existing economic system satisfactory solutions to the problems of deprived areas are not easily found. Indeed, considerable change is necessary in the present priorities for resource allocation, though this point does not always seem to be appreciated. Donnison (1973), for example, proposes that local service centres are established in deprived areas both to improve the services available to people and to act as a focus for local political activity. The centres would combine the functions of a citizens' advice bureau and offices of local government departments, such as housing, social services and education. They might also have library facilities, a planning team, housing aid services, rent officer, probation officer and so on, depending on local circumstances and needs. An area officer would be responsible for the co-ordination of services. He should be a professional of at least the same status as a deputy Chief Officer, a 'company commander', generally responsible for this 'bit of the front' (p. 397). On the political side, local people would be encouraged to take responsibility 'and to nominate their spokesmen to consultative groups attached to statutory services' (p. 398). They must be given information, and offered space in the centres to hold meetings and for other activities. A management committee might consist of local councillors, aldermen, perhaps the local MP, and a few co-opted members. Central government should give grants to local authorities to establish such service centres, and financial support to local groups. Donnison recognises that political activity over the distribution of scarce resources necessarily involves conflict, and suggests that the

> aim is to promote more productive conflict and furnish procedures for successive, temporary arbitrations and agreements. When people conflict over housing, play-space, jobs, welfare services and other things which can in time be extended, subdivided and redistributed almost indefinitely, such arbitration is feasible because everyone can gain something from it. Thus the class conflicts common in urban society are in the longer run relatively benign, unlike the conflicts that arise between religious or ethnic groups, where the contenders are apt to find themselves playing 'zero sum games' in which every gain is someone else's loss (pp. 400-1).

He concedes at the outset that this proposal is 'certainly *not* a solution to the problems of the inner city' (p. 396), but suggests that the value lies in combining aspects of professional bureaucratic activity with political activity, though he does not go on to elaborate how an extension, subdivision and redistribution of scarce resources will come about. It is important to note the implication that some degree of conflict is to be contained within the scope of existing local political institutions. I find this formulation extraordinary, all the more so given current cuts in public spending and a decline in services. Further, it involves a spurious distinction between class conflict, said to be ultimately 'relatively benign' and racial and religious divisions between people, which allegedly are not.

Many studies of poverty and deprivation have highlighted inner-city areas as the, by now, widely acknowledged locus of hardship (Blair, 1973, Ch. 2; Community Development Project, 1977a, 1977c; Cullingford *et al.*, 1975; Department of the Environment, 1977; Donnison and Eversley, 1973; Falk and Martinos, 1975). In recent years successive British governments have financed limited positive discrimination programmes in an attempt to overcome the problems of deprived areas, which it is assumed will also solve the problems of deprived people. There have been General Improvement Areas (1968), Education Priority Areas (1968) and Housing Action Areas (1974) which have all attempted to provide compensatory inputs of a single service, and Community Development

Projects (1969) with a potentially limitless concern with the allocation of services and other resources. In addition there have been several research/action projects like the Inner Cities Projects (1973), the Urban Aid Programme (1968) and the Comprehensive Community Programme (1974), based on the underlying rationale of channelling additional resources to areas of stress. These are 'place-based' attempts to find solutions to the problems of deprivation, attempts which by their very nature have severe limitations, as evidenced by the experience of the USA Poverty Programme of the 1960s (Marris and Rein, 1967), and also in Britain, as acknowledged by Batley (1975), Clark (1972), Flynn (1978), Halsey (1972), McConaghy (1971, 1972) and several reports of the Community Development Projects (1977a, 1977b, 1977c).

In contrast to the thinking behind these government projects and Donnison's proposals, is the argument made out by Ambrose and Colenutt (1975) for radical change in the system of land tenure and land use if planning is to be compensatory and benefit currently disadvantaged groups. They write about the role of private property and the influence of commercial developers in redevelopment schemes in Brighton and London. Their critique is concerned with both the deficiencies of the land-use planning system and with the economic system within which land-use planning operates. Accordingly they consider the adverse effects of commercial redevelopment on local employment opportunities, housing and living costs, on the inequality of wealth, the regressive redistributive effects of redevelopment in general, and the inefficient use of national resources which is often involved. They do not fight shy of the logic of their own analysis in their recommendations for action. These include the restructuring of finance capital and the market in land and property to which it is tied. They urge the replacement of the current land-use planning system 'which responds to supply and demand with a development process which is organised around needs and priorities' (p. 164) and the public ownership of all land and property. In stressing wholesale nationalisation, there is perhaps a danger of putting too great an emphasis on *who* makes decisions, rather than *what* is decided. The nationalisation of land, financial institutions and property will not necessarily enable planning to operate in a positively discriminatory fashion, for this requires the appropriate political will, irrespective of which institution has formal ownership, as these authors recognise. Accordingly, Ambrose and Colenutt (1975, pp. 164-80) also call for a reduction in the power of the property professions, that appointments to the Civil Service and local government be made on political as well as professional criteria, that community action and public participation be increased, and so forth.

A reformist perspective contains an inherent contradiction which is well illustrated by the development of Harvey's work from his earlier liberal formulations to his later socialist formulations. The analysis of urban or social problems these writers present, though not a uniform one, tends to stress the inevitability of poverty and disadvantaged groups under capitalism. The logical prescription for action from this analysis would be radical change, admittedly a severe problem for implementation, whereas many of the actual recommendations and activities tend to be reformist, allowing that there will probably be no major changes in the economic and political system, at least in the foreseeable future. The contrast between this position and the fourth perspective to be considered here partly concerns this issue of reformist solutions. Marxist approaches give an account of urban issues which tie them firmly into the capital accumulation process, and offer an implicit or explicit critique of the exploitation of working people under capitalism. There is a clear integration of political and economic aspects of urbanism, and an associated interest in political activity at national and local level which attempts to radically alter the current distribution of power in society in favour of the working class.

## The Political Economy of Urbanism

There has been a revived interest in Marxism over the past ten years or so, with a re-emergence, re-interpretation and development of the concepts of classical Marxism. Especially in this period of economic crisis, Marxist writers and political activists have been challenged and stimulated to analyse and explain the economic, political and ideological processes at work, to assess the nature and extent of the 'crisis', and to separate out the interrelated aspects of economic crisis and urban crisis. This can be seen as an important corrective to the conventional wisdom which tends to confuse the two and somehow attributes 'the problem' to 'the city', and hence looks for physical and spatial solutions. Thus there have been attempts to develop Marxist concepts as a basis for an analysis of urban situations. This especially applies to urban studies in France, where the government has funded numerous research projects following the 'events' of 1968. This innovative work by Castells (1976, 1977),

Lojkine (1976, 1977) and others can be seen as an important contribution to the development of Marxist thinking specifically in the context of urbanism. Some of it is available in English, associated with the efforts of Pickvance in his capacity as editor, commentator and translator (1974, 1975, 1976, 1977, 1978). Varied contributions by English and American writers include Harvey's later essays (1973, Chs. 5 and 6), work by Broadbent (1975), Friedman (1977), a collection of papers edited by Harloe (1977), papers prepared for the Housing Workshop of the Conference of Socialist Economists (1975, 1976), various conference papers on urban change and conflict (Centre for Environmental Studies, 1975, 1978), and the recently established *International Journal of Urban and Regional Research* (1977 onwards). Much of this work is very recent, explorative in nature and to some extent tentative, often at a very high level of generality, and acknowledged by the authors to be inadequately formulated or as yet incomplete. However, this approach appears to have several inherent strengths not found in the perspectives which I have already discussed. Crucially important, it does not treat economic factors as non-issues. It can take account of power 'exercised' through day-to-day assumptions about how a market economy operates, and it is addressed to the characteristics and purposes of the city development and redevelopment process. One of the fundamental weaknesses of the pluralist and bureaucratic approaches to the study of urban issues and land-use planning is their neglect of economic factors, accepting that economic questions constitute a separate area of study from sociological and political questions and that it is unnecessary to relate these two areas of concern.[5]

Fundamental to writers in this Marxist tradition is the relationship between the development of cities and the capital accumulation process. Lamarche (1976, p. 86), for example, views the city and urban problems as the 'local consequences of capitalist accumulation'. He delineates a specialist capital, property capital, whose primary purpose is to plan and equip space in order to reduce the indirect circulation costs of production. Property capital has a planning role in the way it selects sites, and an equipping role in the types of buildings developed. Its commodity is floor space, let by the square foot, and its profits depend on the difference between construction costs and rent extracted. Property capital mainly caters for commercial, administrative and financial users, rather than providing housing, for example.

> Since he is not a financier, that is to say he seeks something other
> than the average rate of interest on the capital he advances, the
> developer will not operate in the field of housing, unless he can
> convert the advantages provided by the environment into profits
> (p. 96).

The developer will be able to demand a higher price if these environmental advantages are not equally distributed in space, but will only
be interested in luxury residential accommodation as a sideline,
ancillary to larger developments. In order to maximise rental
income, property capital will concentrate its developments in areas
with good situational advantages, favouring high-rise buildings in
such areas, and large developments, where tenants have complementary functions.

Lamarche notes the importance of public investment to
developers, who extract higher rents for sites which are close to
railway stations, government offices, schools, parks and so forth,
though this additional value is dependent upon the activities of
public authorities rather than private developers themselves.
Further, he claims that the development plans of the authorities can
only be realised if they are subordinated to the interests of
developers, and that urban renewal is governed by the priorities of
property capital. Lamarche draws on examples from Canadian
cities. Despite some differences in scope between the urban planning
systems of Britain and Canada, his characterisation shares common
features with some accounts of land-use planning in Britain, as for
example in Counter Information Services' anti-report on housing in
London:

> The history of planning shows that private ownership of land and
> private initiative in, and profit from, its development, set forces
> in play which will always break through the obstacles of any
> planning measures, which start from an acceptance of their legiti-
> macy and permanence (1973, p. 51).

Lojkine (1976, pp. 119-46), like Castells whom I shall consider
next, bases his work on a structuralist reading of Marx following
that of Althusser (1969) and Althusser and Balibar (1970).[6] The
Althusserian concept of the social formation contains four distinct
practices: economic, political, ideological and theoretical, although
the economy is determinant 'in the last instance'. Lojkine argues

that the capitalist mode of production requires a spatial organisation which facilitates the circulation of capital, commodities, information and so on, and that the capitalist city can be seen as a spatial form which, by reducing the indirect costs of production, and costs of circulation and consumption, speeds up the rotation of capital.

> the city is...in no sense an autonomous phenomenon governed by laws of development totally separate from the laws of capitalist accumulation: it cannot be dissociated from the tendency for capitalism to increase the productivity of labour by socialising the general conditions of production — of which urbanisation...is an essential component (pp. 123-4).

Lojkine stresses the tensions and contradictions inherent in capitalism, and hence the capitalist city and land-use planning. Capitalist relations of production, together with modern industry, bring about a growing tendency towards urban concentration and agglomeration. 'Rational, socialized planning of urban development' is necessary to continued capitalist accumulation, but may not be forthcoming due to the existence of three limiting factors (p. 128).

The first problem is that of profitability. Many collective means of consumption — housing, schools, parks, medical facilities, for example — are by their very nature 'opposed to the imperatives of profit'.

> Due to the slow speed of rotation of capital in these sectors, to the risky and discontinuous nature of the progression of demand for them, public transport, schools, research centres, parks...constitute so many domains foreign to capitalist profitability, while necessary to the overall reproduction of capitalist social formations (Lojkine, 1976, p. 133).

Second, competition between firms tends to lead to concentration of investment in some areas, and a widening gap between well equipped and poorly equipped regions, towns or neighbourhoods. Third, there is the problem of the private ownership of land.

It is through state intervention that collective means of consumption exist at all, one of the roles of the state being to maintain the cohesion of the social formation as a whole, which may involve economic as well as legal and ideological measures. Lojkine notes three common characteristics of the urban policies of the developed capitalist states: that state intervention enables the capitalist system to resolve the immediate contradictions that no individual capitalist agent either can or wants to resolve; that it has permitted public financing of unprofitable means of communication and collective means of consumption enabling capitalism to promote simultaneously the development of all the general conditions of production; and that urban planning, while it has had very uneven results, has enabled immediate difficulties to be resolved, as with public health legislation and working-class housing. Attempts at land collectivisation have enabled limited planning experiments to be undertaken, such as the New Towns. According to Lojkine, this intervention does not function solely as a safety valve, but is the reflection of class struggles and worker pressure, obliging the state to limit the spontaneous tendencies of capital accumulation. Certain planning laws seem to be consequences of the 'panic' of the dominant classes, themselves threatened by the 'decay' of the cities.

> But a more thorough historical study shows that behind the apparent non-intervention of the working class, the bourgeoisie's very 'fear' of a possible popular uprising, its general struggle against the development of a revolutionary labour movement, led it to take political measures which cannot be separated from this context (p. 144).[7]

As partial resolutions of immediate difficulties, capitalist urban policies simply carry the contradictions between the necessities of urban socialisation and the necessities of capitalist accumulation to a higher degree. Far from suppressing the class struggle, economic and legal intervention by the state apparatus in the urban domain simply extends its field of application. Lojkine applies this theoretical formulation to two regional studies of urban policy-making, attempting to account for the roles of the state and private actors in large-scale development projects in Paris and Lyons (Lojkine, 1977; Pickvance, 1977).

Castells's work, like that of Lojkine, is based on a structuralist reading of Marx, and for Castells the economic, political and ideological practices of the social formulation are the keys to understanding urban society. 'Urban' refers to 'reproducing labour power', meaning that all aspects of life outside the workplace are bound up in this process. This would include the basic provision of housing, food, heating, clothing, and more broadly the provision of education, health and social services, recreation and leisure-time activities, which train the work-force and maintain its capacity for work. Thus, 'urban units' (the city and the city region) are to the process of reproducing labour power what enterprises are to the productive process, and the 'urban system' is 'the particular way in which the elements of the economic system are articulated within a unit of collective consumption' (1976, p. 153). Analysis of this urban system leads directly to the study of urban politics which can be divided into two constituent elements: urban planning and urban social movements. Urban planning is the intervention of the political system in the economic system at the level of the urban unit, in order to regulate the process of the reproduction of labour power, and the reproduction of the means of production, though 'not every conceivable intervention by M (management) is possible, because it must take place within the limits of the capitalist mode of production, otherwise the system would be shaken rather than regulated' (p. 166). There are two limits: in general there must be no change in the ownership relation, and there must be no direct intervention by management on production. There may be indirect intervention, however, which would include zoning, financial incentives, provision of housing, communications and so on.

Urban social movements, on the other hand, are:

> organisations of the system of actors leading to the production of a qualitatively new effect on the social structure... The term 'qualitative new effect' may refer to either of two basic situations:
> — at the level of *structures*: a change in the structural law of the dominant system (which in the capitalist mode of production is the economic as far as the ownership relation is concerned)
> — at the level of *practices*: a change in the balance of forces in a direction counter to institutionalised social domination, the most characteristic index of which is a substantial change in the system of authority, or in the organisation of counter-domination (1976, p. 151).

Since every social system is in a constant state of change, this throws up 'issues' around which the interaction of actors occurs. Every social issue

> has its more or less immediate source in a contradiction or dislocation between the elements of one system, between elements

belonging to different systems, between these systems taken as wholes, or finally between structures or forms belonging to different modes of production articulated within the same social formation (1976, p. 141).

Thus the study of urban planning is the study of the structures governing the nature of urban society, and of the attempts to resolve or minimise the contradictions or dislocations between aspects of the system. The study of urban movements is the study of the practices of class agents. The first leads to an understanding of the configuration of the urban system, and hence of the social formation; the second to understanding the processes by which it is maintained or transformed. Although this distinction is crucial at the theoretical level, in practice they are 'two sides of the same coin', since structures are articulated practices of class agents, and practices are the interactions between and within social classes as determined by the particular structural configurations of the society.

Having thus set out the conceptual framework, Castells proposes his 'experimental hypothesis'. Since the state expresses at the political level the combined economic and political interests of the dominant classes, then planning cannot be an instrument of social change, but only one of domination, integration and conflict regulation. On the other hand, a process of social change (as distinct from 'reform' which can be achieved by the planning system or within the field of conventional politics) does emerge from the new field of social conflict to which the social movements give expression, when popular mobilisation occurs; when social interests become political wishes, and when new forms of organisation of the reproduction of labour power (e.g. welfare state services) clash with the dominant capitalist ethos of accumulation, competition and profit. Thus, Castells argues, it is the urban social movements and not the planning institutions which are the source of change and innovation within the city. Hence he is interested in organisations which mobilise around particular issues, such as urban renewal in Paris, a proposed motorway in Quebec and squatter settlements in Allende's Chile (1977, Ch. 14). His methodology involves identifying the contradictions, the stakes for each social group intervening, the characteristics of these organisations and an analysis of their activities and effects.

The fourth contribution to be considered here is that of Olives (1976), who utilised Castells's approach in his analysis of local resistance to a proposed urban renewal project in Cité d'Aliarte, a suburb of Paris. Olives concludes that every protest movement springs from the perception of a *stake*, by a *social force*, in this case the attempt to evict immigrant workers, mainly Africans, from their hotels or hostels. The size of the stake must be large enough for urban or political effects to result, but such effects are only possible where there is a minimum degree of organisation of the *social base*. It is neither sufficient for the stake to be large and defined on a social base, nor does the presence of an organisation lacking a social base result in urban or political effects. Legal action undertaken by a purely protest organisation — petitions, letters, requests and so on — can never lead to effective results, because it fails to mobilise people. Given the presence of an organisation, the appropriate type of action and social base, it appears that it is the size of the stake which is decisive in determining whether urban or political effects are obtained, though Olives comments that

> it still remains to be seen to what extent the size of the stake is determined by the real intensity of the contradiction and to what extent by the political conjuncture (balance of forces in the class struggle) at a particular place and time (p. 178).

The conflict situation described by Olives originated in the sphere of housing, but it was not merely an isolated local housing struggle, for the workers involved made links with political groups and hence the housing issue was welded to a class political struggle on other fronts. Olives maintains that such links are vital to the success of the local organisation, otherwise there are two likely outcomes to this kind of protest. The organisation may be dissolved 'through its paternalist integration' into the system, and/or partial reforms may be achieved which tend to preclude any further significant changes being made.

Though the workers in this case successfully resisted eviction and called the whole planning proposal into question, none of their other specific demands, for example rent reductions and lower fares, were satisfied.

> Their political and ideological relevance apart, it is obvious that the latter were out of step with the level of development of the struggles in progress (balance of forces) and the state of development of working class mass organisations (p. 183).

Nevertheless, Olives concludes, the actions of the workers of the Cité d'Aliarte delayed the renewal programme and challenged it in a fundamental way. To carry the struggle to a higher political level needs organisations constituted on a broader basis, affecting the metropolitan area as a whole, and providing linkages to non-urban organisations.

I suggested that this Marxist work appears to have an inherent strength for gaining an understanding of urban issues and land-use planning in its emphasis on the interrelationship between city development and the workings of a market economy. However, it has been subjected to much criticism, varying in scope and intensity. It is important to discuss the criticisms in some detail, to assess how damaging they are. There appear to be six points to be considered here, and in later chapters: the lack of attention given to the characteristics of local organisations; the methodology employed; the claim of universal applicability; a theory of social change; the question of the uniqueness of capitalism; and the nature of the role of the state.

### Characteristics of Local Organisations

Castells focuses on the effects achieved by social movements, and is barely concerned with any other aspects. He is mainly interested in the results of an organisation's activities, not with details of the kind of people involved, their degree of politicisation, the issues which have predisposed them to be active, nor the practical details of their campaigns. In criticism of this emphasis, Pickvance (1976, Ch. 8) stresses the importance of organisational resources in contributing to a group's survival and the achievement of its aims. Of course, any organisation needs resources, both personnel and finance, above some minimum level, to operate at all. Different organisations vary enormously in their resources — money, people's time and energy, information, professional expertise and influential contacts. However, successful groups often have greater resources than unsuccessful groups. If this is so, then Pickvance's criticism of Castells is less telling, for a stress on effects would implicitly incorporate the point about resources. If one focuses on the effects of local action, one is interested to ask whether or not it achieved its objectives, and if not, why not? In practice there is often an evaluation of any particular campaign, perhaps trying to identify if it was wrongly directed; how more people could be contacted, interested and mobilised in support; what further activities might be planned, publicity attracted, and so on.

I would accept that it is necessary to know much more about the characteristics of local organisations than Castells suggests, especially concerning mobilisation — the type of people likely to be active in particular organisations, and the issues and activities likely to attract support. Pickvance, however, tends to over-stress the importance of resources, and to overlook the importance of a group's aims which are related to success and resources. Groups with demands which threaten the *status quo* tend to have fewer resources than those which do not, though they all, with the exception of big business, have fewer resources than the state. Groups with radical demands also tend to be less successful, but I am arguing that this is at least as much to do with the nature of their aims as their relative lack of resources.

In a study of rent strikes associated with the Housing Finance Act 1972, Sklair (1975, p. 269) notes that a 'rent strike is a very complex phenomenon. It has a pre-history, a set of precipitating factors, an organisational and an ideological structure, a case history, and an outcome'. He suggests that the typical rent strike usually involved people who had had some previous experience in a housing campaign. Those he investigated were precipitated by rent increases imposed under the Housing Finance Act 1972, and also linked in people's minds to the conditions of the housing estates, often desperately in need of maintenance. The level of organisation Sklair describes as 'flimsy', based on tenants' associations, many of which were usually only involved in social activities. Some tenants' associations were 'issue-oriented', however, and had experience of organising around a particular grievance. Among the activists, two different ideologies could be distinguished. Some had as one of their objectives the election of a Labour government which would repeal the Housing Finance Act, while others saw the entire system of housing for profit as their target. For these people, electing a Labour government was only the beginning. Sklair gives details of rent strikes in various British cities, and their ultimate collapse. I think that this general approach is a useful one for studying urban protests.

*Methodology*

A criticism of structuralist work put forward by Dunleavy concerns the criteria employed for the selection of issues to study, which he also levelled at pluralist studies as mentioned above:

Castells' paper on the study of social movements is an explication
of how to code or characterize protests once they are selected for
study. But it provides no criteria for the selection procedure itself
and no indication...that a set of latent or non-protests may exist
or have analytical importance (1977b, p. 199).

Dunleavy goes on:

So far as I can see structuralists study virtually any 'struggle'
which is accorded significance in general Althusserian Marxism.
Some British studies on structuralist lines go further and analyze
cause groups whether they use protest tactics or not. Thus Pickvance uses evidence about *a local branch of the United Nations
Association* to derive hypotheses about the role of organization in
urban social movements, while Lambert *et al.* pass directly from a
study of a residents' association involved in the redevelopment
process in Birmingham to a typically structuralist account of
urban conflict, even though the association functioned as no
more than a normal, well-integrated interest group. Such indis-
criminate studies of the micropolitics of the city cannot sub-
stantiate structuralist hypotheses.

A study of protest which is to be more than theoretically
suggestive and which is to transcend the intrinsic interest of a
particular instance of protest must illuminate a fundamental
power relationship within the urban system, particularly one
usually tending to keep an issue latent (p. 200).

His own study of local community action over the issue of mass high-
rise housing is formulated in the light of these criticisms (Dunleavy,
1977a, 1977b), First, he selected a topic which had not surfaced in
conventional politics, namely the building of high-rise housing for
local authorities, despite its unpopularity with tenants and prospective tenants. He went on to explore the relationship between
specific local authorities, central government policy for financing
council housing and the role of large construction firms, which were
influential in getting comprehensive high-rise housing schemes
accepted, allowing them to use systems building technology.

The activity of local organisations needs to be put in the context of
more general inactivity, in both the spheres of production and
consumption: in the workplace and in the community. I argued

above that routine assumptions about power operate constantly to keep issues latent, and to restrict the scope of debate and decision-making to relatively 'safe' questions. This happens in several ways: the issues are obscured, either consciously or unconsciously; people have inadequate and fragmented scraps of information which have to be pieced together for an understanding of the full picture; there often seems no point in contesting issues which appear not to be negotiable. On this point Dunleavy (1976, pp. 430-1) argues that disfavoured participants may protest at the limits imposed by gate-keeping activity, but they accept and operate within the limits set by the issue system's boundaries because they rarely have the power to alter them. Hence, while focusing on the activities of local organisations, it is important to note the limited extent of this activity.

### The Claim of Universal Applicability

Structuralist writers claim universal applicability for their formulation, criticised by Pahl (1975, pp. 280-4), who argues for the uniqueness of particular capitalist countries, based on their different historical development and local variations in current practice. Pahl stresses the need to look at British capitalism specifically, without unquestioningly assuming that models which derive from French experience will be appropriate.

I am concerned to understand land-use planning in a capitalist economy, and have argued above that work in the Marxist tradition appears to have several strengths for a study of urban issues. The fact that some of this work has been developed in France should not render it immediately valueless for a study of land-use planning in Britain. The difficulty of making valid generalisations arising out of the uniqueness of particular societies and settlements is acknowledged by Elliott and McCrone (1975, p. 36) and again by Pahl (1975, p. 273), who claims that 'since different groups benefit at different times in different parts of the same city, common city or nationwide situations of deprivation rarely occur.' There is a sense in which every settlement is unique, just as every person is unique. Settlements, like individuals, have their own specific histories, though having said this, it must be acknowledged that individuals are part of wider societies and settlements are part of the nation-state and are affected by general economic conditions, the law of the land and the general political situation. It is the focus on settlements that is not helpful, and this discussion requires the introduction of the concept of class, for deprivation is not random, neither socially nor spatially, as Pahl himself acknowledges elsewhere (1975, Parts 2 and 3). Indeed, I considered this relationship between social status, type of housing, quality of environment, accessibility to facilities and so on in Chapter 1.

*A Theory of Social Change*

Pahl widens this critique of universal applicability and criticises the structuralist perspective as a theory of social change, claiming that there is little evidence from British experience to support a view of urban social movements as leading to radical change in British society. To illustrate this position he takes two issues which one might expect to have elicited a general reaction from local groups and consequently co-ordinated and cohesive activity across local authority areas. These are the 1970-4 Conservative government's Housing Finance Act 1972, and the more general issue of property speculation which has been particularly prevalent from the mid-1960s to the early 1970s in many British towns, though notably in London. Neither of these issues provoked collective organisation on a national basis, however. Many attempts on the part of tenants' associations to organise against the Housing Finance Act were not successful (Sklair, 1975, pp. 250-92). At parliamentary level the Bill was contested in committee, and subsequently some individual local authorities attempted to salvage a limited advantage by negotiating rent increases below the level specified in the Act. Sklair shows that the protesting authorities, and the local authority tenants who engaged in rent strikes, were characterised by their isolation and lack of support. Similarly, the activities of property speculators did not stimulate working-class collective action against the private ownership of urban land although they did stimulate many working-class organisations to campaign at local level, but it was industrialists, according to Pahl (1975, p. 274), who apparently urged the government to take action and hence to adjudicate the conflict of interests between finance capital and industrial capital:

> Controlling the excess profits of property speculators became a national political issue at the end of 1973, when, amongst others, Lord Plowden, Chairman of Tube Investments, one of the largest of British industrial enterprises, wrote to *The Times* urging government action. It is significant that this pressure to take action seemed to come at least as much from the controllers of industry as from trade unionists, and was directed, evidently,

against the capitalist system in housing and land and not against the capitalist system in industry.

A further criticism of Castells concerns his use of the term 'collective consumption' — said by Pahl (1977, pp. 166-70) to be inconsistent and confusing — and his emphasis on the consumption process one-sided and misleading. A restricted focus on urban social movements overlooks working-class organisation and trade union bargaining in the sphere of production, yet the increased wage levels which are the outcome of this negotiation affect consumption patterns. In contrast to this, Pahl claims that urban social movements have very little effect on consumption. Similarly, in times of economic stagnation an erosion of buying power due to inflation and wage restraint also affects consumption patterns, this time adversely. The whole issue of the link between 'the struggle in the community' and 'the struggle at work' is extremely important, involving problems of sustaining and co-ordinating locally based action. It is true to say that activity in local organisation is a minority activity, especially where radical demands are involved. Moreover, as Pahl points out, such activity as there is is uncoordinated either regionally or nationally. Co-ordination on a London-wide basis is illustrated by the campaign against the London Motorway Plan (Hillman, 1971; Thompson, 1977) and the campaign against the Community Land Bill (Land Campaign Working Party, 1975).

*The Question of the Uniqueness of Capitalism*

A final criticism of Castells and others by Pahl (1975, 1977) concerns the stress laid on the uniqueness of capitalism and their view of urban issues and problems in capitalist societies being a product of *capitalism* specifically. He maintains that useful insights can be gained from comparative studies, taking into account urban problems in Eastern Europe, for example. I have acknowledged above that problems of resource allocation are not unique to the capitalist mode of production, but I am not prepared to accept that these problems have exactly the same causes in centrally planned and market economies. Hence I am not attributing them to large-scale industrial organisation *per se*. I argued in Chapter 1 that Britain has a capitalist mode of production, and hope to show that it is this mode of production specifically which creates particular contradictions in land-use planning in Britain. Thus, irrespective of the fact that there may be some apparently similar problems in cities in centrally planned and market economies, I am arguing that it is the contradictions of British capitalism which account for the contradictions of land-use planning. Thus it is the existence of private ownership of land and capital and private initiative in development which are at the root of the problems of resource allocation in Britain.

These several criticisms may be thought to call into question the usefulness of this Marxist approach, though some critics want to retain and utilise some of the concepts. Dunleavy (1977b, p. 193), for example, suggests that this structuralist work has 'fundamentally redrawn the lines of debate in the study of urban politics', though his own study aims to overcome what he considers to be two weaknesses, involving

> a modification of structuralism to take account of two central points made by neo-elitist writers, the need to recognize the capacity of the state apparatus and other dominant groups to suppress or deflect hostile political developments, and the consequent analytic necessity of studying political inactivity or quiescence in addition to overt political action (p. 194).

An understanding of the land-use planning process in Britain requires a theoretical perspective which can take account of the economic, political, spatial and social questions involved, and can go beyond the one-sidedness or restricted focus of the pluralist and bureaucratic approaches discussed in the first part of this chapter. I consider the general view of the city as the outcome of the capital accumulation process to be a useful one. Similarly, the concepts of stake, social base, social force, protest effects and so forth developed by Castells and Olives in their analysis of urban social movements appear to be a useful starting-point for conceptualising the activities and structure of protest groups. In the foregoing discussion of the criticisms of Castells's and others' work I am not convinced by the arguments concerning the claim of universal applicability and the uniqueness of capitalism, as put forward by Pahl, though I do accept that it is necessary to know more about the characteristics of local protest organisations and the role played by the local authority, to locate activity within the context of more general inactivity, and to consider critically Castells's and Olives's formulation as a theory of social change. I have not yet considered the role of the state, as characterised by these structuralist writers, and this will be incorporated in the following chapter in a more detailed discussion of the organisational context within which land-use planning operates.

The theoretical perspectives outlined here all focus on different aspects of the distribution of power in society, and have various strengths and weaknesses. These approaches each have their own specific emphasis, and allow one to focus on particular aspects at the expense of others. Thus the pluralist view, for example, stresses overt political activity, but does not penetrate beneath the surface to less obvious or concealed assumptions of power; the bureaucratic view stresses the power of public bureaucracies, but ignores the role of business interests and their influence on government policy and so on. In the next three chapters I shall amplify the theoretical approaches outlined here in connection with three themes: the government context of land-use planning, the role of professional planners, and the scope for public involvement in planning decision-making.

## Notes

1. Researching power relationships may be very difficult. To conclude that power is being exercised one needs to show that the individual or organisation would have acted differently, and has been constrained from so doing. Non-power factors such as coincidence, changes of mind and misunderstandings need to be allowed for. There are problems of reliable information, identifying all the significant 'actors', assessing their roles, and the nature of the interrelationships between the various parties. See, for example, Domhoff, 1978; Dunleavy, 1976; Lukes, 1974, Ch. 8; Parenti, 1970; and Wright 1978, pp. 11-14, Ch. 3.

2. One of the main issues to be raised in connection with the alleged equality of the various competing interest groups concerns the 'power of the unions' thesis. Though this is only indirectly related to my main theme, it is important to mention it, for it is a crucial aspect of the distribution of power in contemporary Britain. Several writers have argued against the pluralist view of labour and management competing on equal terms. See, for example, Blackburn and Cockburn, 1967; Fox, 1973; Friedman, 1977; Miliband, 1973; and Minkin, 1977.

3. The term 'community' has been defined in very many ways, and subject to much consideration and debate. Different aspects are stressed, for example by Bell and Newby, 1971; Dennis, 1968; Frankenberg, 1966; and Rose and Hanmer, 1975.

4. Central Statistical Office: national income and expenditure figures. Broadbent (1977, Ch. 2) discusses different ways of estimating state expenditure, pointing out the scope for possible 'double-counting'. He suggests that these figures are an over-estimate of state expenditure as they include money which is merely being redistributed by the state in the form of welfare payments and suchlike. His own estimate of the size of the public sector is approximately 30 per cent of GNP, represented by the purchase of goods and services (p. 57).

5. Urban economics sees city organisation in terms of competitive bidding for sites, where land values reflect demand. At equilibrium, rent levels are said to be equal to the marginal productivity or utility of the land. This view is exemplified by Alonso

(1964) and Muth (1968).

6. Though Althusser himself rejected the label 'structuralist'. Structuralist theories have been developed in other disciplines — linguistics, psychoanalysis, psychiatry and anthropology. General references are de George and Fernande, 1972; Lane, 1970; and Robey 1973.

7. See also Glass (1955) for a discussion of the British tradition of anti-urbanism, and the retreat of the middle classes from industrial towns. 'unhealthy and hideous', and identified with the working class. This concentration of working people was much feared, according to Glass, for it 'represented a formidable threat to the established social order' (p. 16). Howard's classic *Garden Cities of Tomorrow* (1902) should also be noted in this connection, with its original title: *Tomorrow: a Peaceful Path to Real Reform*.

# 3  THE ORGANISATIONAL CONTEXT OF URBAN PLANNING

This chapter takes up one of the themes raised in the first chapter — the organisational context within which land-use planning is carried on — and explores it in relation to the four theoretical perspectives outlined in Chapter 2. The different theoretical approaches do not all lay the same stress on this particular issue, so my treatment varies considerably in length and detail. There are two aspects of this organisational context which can be distinguished. The one concerns planning as a government activity, and the other concerns the development of land-use planning as a professional activity and academic discipline. To some extent there is a tension between the demands arising out of these two contexts — the administrative and the professional — as I hope to show, both here and in the following chapter. I shall turn to look at the professional context first.

## Planning as a Professional Activity and Academic Discipline

Since the early years of this century British planning has developed as a professional activity, dedicated to a systematic ordering of land uses. Within professional circles it has become common to put forward a view of planning as rational problem-solving, where future targets are spelled out and various means for achieving these objectives considered and assessed.[1] This view was originally taken from theories of business management, particularly American work of the late 1950s and early 1960s (Davidoff and Reiner, 1962; Dyckman, 1961; Lindblom, 1959) and applied to the British context (Chadwick, 1971; Friend and Jessop, 1969; Friend, Power and Yewlett, 1974; Greenwood and Stewart, 1974; Stewart *et al.*, 1968). The main theme is that planners should act in a non-arbitrary manner, making judgements explicit, based on reason, and that their decisions should be backed by adequate information, quantifiable where possible. To take a simple example, a decision to rehabilitate dilapidated houses rather than to demolish them and rebuild should be based on a consideration of the advantages and disadvantages of these two possibilities. This might include the relative costs, the expected life of the renewed housing, the relative standards of accommodation that could be provided, the advantages and disadvantages of the physical form the housing would take, the costs and benefits to existing and prospective tenants or owners, the ease of doing the work and the kind of building firms which might be employed and so on. Of course very many planning problems are more complex than this example: how to attract industry to the depressed North, or London's Docklands; whether a proposed new airport will have positive spin-off in terms of stimulating the local economy; traffic management in a city the size of London and so forth. But the value of decision-making techniques lies in their capacity to enable one to break down complex problems, and articulate clearly the various possibilities and their likely implications. As Roberts (1974, pp. 50-1) puts it:

> Decision theory provides a highly appropriate conceptual framework for land use planning and helps to clarify both requirements for the process as a whole and for its constituent parts. It emphasises the choosing nature of making decisions, the necessity to clarify options, to value the possible outcomes and to blend such values with the probability of the option's occurring. The fact that it is impossible to guarantee the future becomes clear, also that even the 'right' decision depends largely on which kind of strategy is preferred. Uncertainty thus becomes a recognised major component of probabilities, but cannot be eliminated, and the right decision is recognised as totally dependent on the values attached to the various outcomes.

This approach gives three broad stages to the planning process, which is generally viewed as a cyclical and reiterative one. First there is the identification of goals, second the means of achieving such goals, and finally administration, implementation and evaluation of preferred programmes (Roberts, 1974, Chs. 7 and 8). For example, the planning goal for a blighted inner-city area might be to put the land to some profitable or worthwhile use to help to rejuvenate a depressed local economy. The encouragement of any one of a whole range of different kinds of development — industrial, commercial, residential, community uses such as a sports hall or a park — would constitute different means of achieving this objective. The planning authority would need to decide between them, or to have some idea of its priorities — for example, whether it preferred to see offices rather than housing, or industrial development which would provide industrial jobs rather than commercial development which would generate a demand for office workers. Accordingly, the planning authority would give planning permission for proposed development which appeared to be in line with its overall aim. After the chosen development had taken place, the authority should then monitor the situation and assess how successful it had been in meeting its declared objectives.

Knowledge of all the factors which might be thought to have a bearing on a particular issue will be limited, as recognised by Davidoff and Reiner (1962), for example. They argued that the goals of the planning process are debatable, hence there cannot properly be said to be *correct* planning decisions. The objective is to have some confidence that the choice made 'was at least as reasonable or more reasonable than any other alternative' (in Faludi, 1973, p. 28). Ways of eliciting both facts and values include market research, public opinion polls, public meetings, interviews with 'informed leaders' (in Faludi, 1973, pp. 28-9) and suchlike, though these authors suggested that the methods available at the time of writing were far from adequate, making a plea for improved techniques for determining ends, and relating ends to means. They urged that planners be taught such techniques as part of their professional education.

Indeed this call for methodology, techniques and useful contributions to the planners' toolkit has been taken up by many, both writers and practitioners, with the aim of making planning a more exact, quantifiable and, some would say, scientific activity.[2] The model here, either implicitly or explicitly, is natural science, and it is worth commenting briefly on its appeal. The attractiveness of scientific method and activity for planners would appear to be twofold. First, the aim of science is to discover and understand the nature of our physical environment with a view to controlling it. Since the declared objective of land-use planning is some better ordering of resources and land uses, there is an obvious similarity in approach. Second, there is respectable 'scientific' status for planning practitioners.

Competing philosophies of science (e.g. Kuhn, 1962; Lakatos, 1970; and Popper, 1959, 1963) give various accounts of proper scientific method, the processes by which scientific knowledge is acquired, and logical structure of the products of scientific research. Scientific method usually involves repeated observations, precise measurement, controlled experiments under laboratory conditions, testing hypotheses and so on. Implicitly there is the assumption of science as a neutral activity, objective and value-free, purely concerned with establishing and demonstrating physical and chemical relationships. This assumption is strongly challenged by Rose and Rose (1976a, 1976b), for example, who ask 'Whose science is it?', 'Who pays for it?' 'Who decides what is researched?' and 'Who benefits from it?' They point to armaments research, the development of chemical and biological warfare and the vast resources devoted to the space programme, and argue that science is institutionalised in research departments and meshed into the machinery of the state through the medium of research grants. Projects of some potential interest to the state attract funding, while others, which might be of equal or greater interest intellectually, and morally more valid, do not. Harvey (1978) makes a similar point in discussing the neo-Malthusian view of the relationship between population and natural resources, which argues that without severe limitations on population the earth's natural resources will soon be exhausted, with chaos ensuing. He also argues that institutionalised science is not neutral, but serves specific interests.[3]

However, science is the name of the most respectable kind of knowledge in Western societies and, in the words of the Roses, is often considered to be 'the only valid way of understanding and apprehending the universe' (1976, 1976a, p. xxiv).[4] Physics is the model to which all science aspires, and thus to

> mathematise, to formalise, becomes the hallmark of the mature hard science against the immature soft science. Nor is this only an issue in the natural sciences, as physics becomes the model for all human knowledge, and what cannot be encompassed by its mode of rationality is illegitimate (Rose and Rose, 1976, p. xxv).

There has been a considerable response within planning to this call for mathematisation, especially with the development and increasing availability of computer technology, so that there is a growing range of statistical models, forecasting devices and aids to decision-making. Roberts (1974) gives an account of these, differentiating between techniques to aid the planning process and techniques for providing information planners need as inputs into that process. An example of the first kind is critical path analysis, for planning out how a project will be implemented, the order of the various steps, who is to undertake them, what resources are required at each stage and so on. Another is cost-benefit analysis, where various planning possibilities are evaluated in terms of the likely costs and benefits to different sections of the community which are expected to result, were the options implemented. The most ambitious 'technique' involves comprehensive co-ordination of all local authority services, and goes by the title Planning, Programming and Budgeting Systems (PPBS). The organisation needs to work out its overall objectives and then to organise its staff and financial resources accordingly, allocating priorities to various activities. There is an ongoing assessment and evaluation of the success in achieving the specified objectives, and subsequent reorganisation or reallocation of resources where necessary. Techniques of the second kind include population forecasting, forecasting housing need, job supply, calculating the viability of shopping centres, as with gravity models, and the environmental capacity of different areas, such as residential neighbourhoods, urban parks, countryside and sea-shore.

An enthusiasm for techniques, uncritically held, can have serious consequences, for it effectively limits consideration to those issues which are amenable to arithmetic, and can be reduced to numbers — though not always meaningfully. Dennis (1972, Chapters 11 and 12), for example, argues very effectively against the accuracy and relevance of housing condition surveys carried out in connection with planning decisions over the future of the Millfield area of Sunderland, where number values were given to the various household amenities and scores added to produce some overall numerical indicator of housing fitness. Similarly, in public participation exercises there is a problem of adding together different opinions on a particular proposal. The use of arithmetic involves the assumption that all opinions carry equal weight, and that the various individuals, organisations or firms are somehow of equal standing. An example of this from the London Borough of Southwark's consultation programme over the North Southwark Local Plan serves to illustrate this.[5] Opinions were sought from local residents' groups, churches, primary and secondary schools, individuals and organisations with a 'wider interest', including transport operators, Trades Councils, the Council of Voluntary Service, social and health agencies, further education establishments, local employers and industrialists. The 'major problems' had been identified in advance by the Planning Department under the headings: population and housing, character of the area, transport, employment and industry, community services, other facilities. Residents' associations put high priority on the need for more housing, particularly for families, more industrial employment suited to the skills of local people, and improved shopping facilities. Also important to them was the lack of open space and inadequate public transport. Employers were concerned about the very run-down environment, poor facilities available for workers, including a lack of housing for sale. They criticised the low level of investment in the area and its inadequate road network. Transport operators stressed the poor public transport service, road congestion and inadequate road network. The difficult thing for the local authority is to take account of all these varying opinions and decide which aspects merit priority, for it is not meaningful to try to add them arithmetically. There are similar difficulties with cost-benefit analysis, where one needs to be able to attach a monetary or numerical value to relatively intangible and subjectively perceived qualities such as the importance of historic buildings, open space, peace and quiet and traffic congestion (Lichfield, 1960; Pearce, 1978; Walters, 1972).

There are both conceptual and practical problems with this view of planning as rational decision-making which no amount of technique can make up for. Roberts (1974, Preface) is careful to point out that techniques are only a means to an end, that of making planning serve people better, and Broadbent (1977, Chs. 5 and 6) makes a similar point. Others have been less careful in this respect and have urged what might be seen as a retreat into pseudo-science, mystification, lengthy time-consuming 'data-gathering' exercises. Mathematical models, like computer programmes, are only as good as their assumptions, some of which often seem trivial, and hardly justifying the complicated procedures which accompany them. Months of painstaking work may be involved before the model yields results. There is a tendency to become taken over by the whole process, and in one's absorption to lose sight of the very limited assumptions and parameters. I am not arguing here against being informed, or that technical calculations have no place. Clearly, they have. What I am arguing against is this call for mathematisation which is often spurious, in an attempt to make planning something that it is not. The fundamental choices and decisions are political ones, and no amount of formulae, graphical representation, cost-benefit analysis and suchlike will change this. Matters of value, the weighting to be accorded to them, the resolution of differences of opinion and interpretation do not lend themselves to quantification.

Yet this stress on quantifiable data has the effect of downgrading qualitative information as 'subjective' or 'political' and hence of questionable validity. Facts are said to speak for themselves. But what we are often concerned with in planning are values and opinions.

Associated with this pursuit of rational decision-making and planning methodology is the conceptualisation of the city or the firm as a system, comprising a series of interrelated parts, such that any action or decision may have consequences and effects, intended or otherwise, throughout the network. Obvious examples of interrelationships between parts of the urban network come to mind. The expansion of residential suburbs, while jobs are concentrated in a central business district, will increase the pressure on the transport system as an increasing number of commuters will have to travel longer and longer distances each day to work. Another example is provided by the services available in a declining residential area. As people move out of the area, those who remain are less and less able to support the shops, schools and other social facilities. Thus a spiral of decline is started, where shops close and the level of services gradually deteriorates still further, causing more people to want to leave.

A systems view is clearly and explicitly spelled out by Chadwick (1966, 1971) and McLoughlin (1969), for example. This approach has a basis in functionalist theory, first developed by anthropologists such as Radcliffe-Brown and Malinowski as an explanation of the allegedly static, unchanged nature of certain 'primitive' societies (Cohen, 1968, Ch. 3; Coser and Rosenberg, 1964, pp. 629-50), and the structural-functionalism of sociology, particularly associated with the work of Parsons (1949, 1951). The stress is on the stability of social wholes and the interconnectedness of parts, so that the function of a particular social institution — such as marriage, education, law or social stratification — is seen as the contribution it makes to the total functioning of the social system. Cohen (1968, Ch. 3), Goddard (1972), Rex (1961, Ch. 4) and Simmie (1974, Ch. 2) all give critical appraisals of functionalism, and it is not my purpose to repeat these here. What is of interest at this point is the way in which land-use planning has adopted elements of the functionalist perspective in systems theory, as an ostensible aid to understanding and managing the 'urban system', where it is

assumed that the processes of city life are systematically related;

and given sufficient information these relationships can be identified; and that town planning is concerned with the rational manipulation of these identified processes, with a view to achieving more desirable ends than those which arise without the interference of town planners (Simmie, 1974, p. 33).

There are several attractive aspects of a systems view. The emphasis placed on the interrelatedness between parts is confirmed by everyday experience, as for example in the political, economic and ecological interdependence between nations, power blocs, trade associations, natural processes and so on. The supposedly comprehensive scope of a systems view thus seems highly appropriate in the face of the multitudinous interconnections between different aspects and parts of the world, the nation, the city and the organisation, and different facets of human activity within these various contexts. A second advantage, claimed by McLoughlin, is that a systems view provides common ground and a common language, welding together professionals from different disciplines — geographers, economists, social scientists, architects, engineers, surveyors — such that their distinctive contributions can fit into the wider framework of the system to be planned and managed. Third, he argues that conceiving the world as a system will induce a greater degree of humility in planners, for

> as we probe deeper and deeper into the systems we treat, we become increasingly aware of the labyrinthine nature of human motivations, choices and actions. We realise too that the physical planner's unique skill is in the manipulation of only the spatial element of life — a small facet of the kaleidoscope of human existence. Even then, with systems so large and complex, if we can raise their performance levels by but a few degrees we can congratulate ourselves (p. 312).

And fourth, the systems view offers what seems like a precise methodology, neatly dovetailing in with the development of mathematical technique and scientific method already noted. As Chadwick remarks, perhaps with some relief, the systems view of planning 'gives a coherence and philosophically satisfying core to a former incoherent discipline' (1966, p. 185).

Aspects of the systems view have now become so widely held among British planners that the theory has the status of conventional wisdom, though McDougall (1973) and Simmie (1974), for example, argue that this generalised acceptance is quite unmerited. McDougall's view of planning is of a particular form of social decision-taking, which needs to be considered and understood in its structural and ideological context, for decisions about spatial distribution are taken within the social, economic and political structure. She argues that the 'solution is not to "depoliticise" planning in an attempt to create a "scientifically" based planning relying on interest-free knowledge, for this is not possible' (p. 83). Both Chadwick and McLoughlin, by contrast, abstract the ecosystem from its social context. Goal formulation, for McDougall, is the most important part of the planning process, concerned with the establishment of value priorities — essentially political questions. Thus goals cannot be derived from scientific method, which is concerned with the establishment of uniformities in empirical phenomena. Further, systems theory 'dehumanises' people, viewing them as mechanical objects to be controlled by 'guidance systems' rather than as conscious, purposive individuals who can choose to create and change their social institutions.

Both McDougall and Simmie argue that 'systemness' cannot be assumed, but must be discovered, yet McLoughlin and Chadwick make such an assumption and proceed to transfer the terminology of general systems theory — equilibrium, information, entropy — and to apply it to land-use planning.[6]. There is a serious problem of defining the supposed system boundaries, which has given rise to a distinction being drawn between 'open' and 'closed' systems, where it is not a requirement of the former that boundaries are precisely defined, nor inputs and outputs across the boundary clearly specified. Cities are too complex to allow straightforward boundary definition, and it is an impossible task to catalogue all the implications and ramifications of altering any relatively large component. Administrative boundaries do not help here, as many people who live beyond a city boundary commute to work each day and are part of the city as an economic unit. Investment decisions may be made miles away from a particular city, yet affect it, often very seriously. Seemingly 'local' issues all raise questions about national policies for investment, and ultimately the nation's command over international resources, as pointed out by Mellor (1975), for example. Simmie also argues that there is an inherent conservatism of both structural-functionalism and systems theory, and this notwithstanding the stress put on dynamic processes. These theories assume a high degree of unanimity on values and aspirations. As such, according to Simmie, they cannot account for the possibility that social inter-action may be characterised by conflict rather than co-operation, and further, they cannot account for change, because there is an assumption that people will choose to co-operate to ensure the survival of the group. Simmie's own work is concerned to argue just the opposite: that there is a lack of consensus on values in modern industrial societies, and an inherent conflict over scarce resources.

Thus the adoption of mathematical decision-making techniques and the conceptualisation of the city as an 'urban system' have had an important influence on the development of land-use planning as an academic discipline and professional activity. I argued above that the limitations to planning powers in Britain are many and serious, and this in itself has been a factor in the development of increasingly complex and sophisticated techniques, where planners have either unwittingly thought to become more powerful through the applica-tion of a more rigorous and precise methodology, or tended to divert their attention towards techniques as a compensation for relative powerlessness, as argued by Broadbent (1977, p. 213), for example. Yet this would-be science is practised in a political context, as an activity of government, for unlike professions such as law, medicine or engineering, it is virtually impossible to work as a planner in Britain outside the public sector. Hence it is important to consider some aspects of this administrative context within which land-use planning operates.

**The Administrative Context of Planning**

This is the focus of interest of those theorists mentioned above who are concerned with the power of government bureaucracy. Amongst other things, they stress the expansion and increasing complexity of local and central government administration. Indeed, the growth of large bureaucracies has been a significant feature of the twentieth century, first described systematically by the German sociologist Max Weber (see Gerth and Mills, 1946; Coser and Rosenberg, 1964, pp. 465-73). He was convinced of the efficiency and rationality of this kind of organisation, with its departmental specialism and hierarchical ordering of posts, and correctly predicted that the development of bureaucracy which he observed roughly between 1890 and 1920 would become very much more generalised.[7] Increasing numbers of people are now employed in the service sector of the economy, rather than in the primary producing or secondary manufacturing sectors. Very many of these are involved in both public and private bureaucracies: in central and local government departments, other government-financed organisations like the BBC, the Manpower Services Commission and suchlike, the administrative side of nationalised industries, and private commercial firms.

**Table 3.1:  The Labour Force, UK, 1967 and 1977**

|  | 1967 Thousands | Per Cent | 1977 Thousands | Per Cent |
|---|---|---|---|---|
| Total working population | 25,490 |  | 26,327 |  |
| males | 16,704 | 65.5 | 16,261 | 61.7 |
| females | 8,785 | 34.5 | 10,066 | 38.2 |
| Employees in production industry | 10,854 | 42.6 | 9,318 | 35.4 |
| Total, manufacturing employment | 8,319 | 32.6 | 7,352 | 27.9 |
| Financial, business, professional, scientific services | 3,531 | 13.9 | 4,773 | 18.1 |
| National and local government service | 1,471 | 5.8 | 1,629 | 6.2 |

*Source: Annual Abstract of Statistics (HMSO, London, 1977 and 1978).*

Table 3.1 shows that for the ten-year period 1967 to 1977 the total working population increased slightly by 3.3 per cent, the total number of employees in productive industry fell from 42.6 to 35.4 per cent, and those involved in manufacturing from 32.6 to 27.9 per cent of the total labour force. Both financial, business, professional and scientific services and national and local government services showed increases over the same period, of 13.9 to 18.1 per cent and 5.8 to 6.2 per cent respectively. This amounts to a spectacular rate of growth in financial, business, professional and scientific services of 35.1 per cent, and a relatively high rate in national and local government service of 10.7 per cent, compared to the 3.3 per cent increase in the total working population, and this despite cuts in government expenditure which have affected expansion and staffing levels.

This table gives a misleading impression of the number of people employed by government, however, many of whom work in industry or professional and scientific services. Hence it should be read in conjunction with Table 3.2, which shows the total number of public-sector employees rising from 23.9 per cent of the work-force in 1961 to 29.6 per cent in 1976. Employment by local authorities shows the largest increase over this period from 7.6 to 12.2 per cent. Central government employees (excluding the armed forces) also show an increase from 5.3 to 8.1 per cent.

**Table 3.2: Employment by Sector, UK, 1961-76 (per cent)**

|  | 1961 | 1971 | 1976[a] |
|---|---|---|---|
| Central government |  |  |  |
| HM Forces and Women's Services | 1.9 | 1.5 | 1.4 |
| Civilians | 5.3 | 6.4 | 8.1 |
| Total | 7.3 | 8.0 | 9.5 |
| Local authorities |  |  |  |
| Total | 7.6 | 10.9 | 12.2 |
| Public corporations |  |  |  |
| Total | 9.0 | 8.2 | 7.9 |
| All in public sector | 23.9 | 27.1 | 29.6 |
| Private sector |  |  |  |
| Employees | 68.9 | 65.1 | 62.8 |
| Employers and self-employed | 7.2 | 7.8 | 7.6 |
| Total | 76.1 | 72.9 | 70.4 |
| Total employed labour force | 100.0 | 100.0 | 100.0 |

*Note: a. provisional*
*Source: Social Trends (HMSO, London 1977), Table 5.7*

The growth in the scope of activities and budgets of central and local government to which I referred in the previous chapters has thus been paralleled by a corresponding expansion in personnel. During the 1960s there was an emerging awareness of the inadequacy of the administrative structure in the face of the increasing tasks government was taking upon itself, and a number of investigations were carried out recommending a variety of changes in the administrative structure and procedure and methods of management at both central and local government levels. These include reports on the Civil Service (HMSO, 1968a), the management of local government (HMSO, 1967a), the staffing of local government (HMSO, 1967b), local authority management structure (HMSO, 1972a), the co-ordination of the personal social services (HMSO, 1968b) and the reorganisation of central government departments (HMSO, 1970). Common to all these reports is the recommendation that the administrative structure be overhauled to allow for the provision of a more comprehensive approach, together with the adoption of modern business management techniques, originally developed for use in private firms.

Details of these processes in the Civil Service are given by Garrett (1972), who puts it this way:

> The Service has been exhorted, mostly by young politicians and elderly industrialists, to make itself more 'business-like' and teams of business men have been hired to range through departments in search of areas in which to apply business methods (p. 7).

Cockburn (1977, pp. 6-24) traces the adoption of decision-making techniques and corporate management from their origins in business enterprise, and argues that it 'was a response to similar kinds of problems to those corporate management responded to in the business world: growing size and complexity of activities, the need for financial stringency' (p. 17). Management courses were developed in universities and polytechnics, and management consultants also played a role in disseminating new methods:

> management consultants were in most cases the same firms who were earning their bread-and-butter modernising the management systems of industry and commerce. They were thus clearly the nearest thing to a direct channel of ideas and methods from the business world to the local state (p. 21).

The Fulton Committee on the Civil Service (HMSO, 1968a) made a number of recommendations for major reforms based on the Committee's appreciation of the shortcomings of the existing structure. It declared the Service to be inadequate in a number of respects, being essentially based on the philosophy of the amateur all-rounder, having a system of administrative classes where administrators took precedence over scientists, engineers and other specialists, and with too few civil servants being skilled managers. Changes in structure and procedure were proposed to meet each of these objections, as discussed by Garrett, for example. Similar changes also affected local government, particularly the adoption of some version of corporate management (Cockburn, 1977, Ch. 1; Darke and Walker, 1977; Knowles, 1977; Richards, 1975). The purpose of this management approach was to allow local authorities to act as a unified co-ordinated whole, overcoming the fragmentation of departmentalism to some extent. The recognition of the interrelationships between social and economic activities, and hence the activities of government departments, arose, at least in part, out of an awareness of inter-linkages provided by a systems approach to urban structure. This resulted in the adoption, either wholly, or more usually partially, of a co-ordinated management structure, to bring together closely related services. In practice there are differences between authorities as regards both the administrative structure and procedures chosen (Knowles, 1977, Ch. 4), but essentially what is involved is the creation of a central Policy Committee, comprising the Chief Executive Officer and Chief Officers, and key committee chairmen. The corporate planning process involves the setting of broad objectives for the authority's service provision, making strategic policy choices, assessing the means available, the resources and constraints, and ideally an ongoing evaluation of performance. I mentioned this general co-ordinating approach above in connection with Planning, Programming, and Budgeting Services. Other techniques as aids to management such as critical path analysis, threshold analysis, goals achievement matrix and so on should also be noted here (Roberts, 1974).

The adoption of corporate planning is justified as an attempt to achieve greater efficiency, both of the use of local authority resources — staff, finance, land, buildings and equipment — and in providing comprehensive services suited to the needs of the people living within the authority's area. It is a complex process, involving numerous co-ordinating meetings and seminars, the writing of discussion papers to be circulated, amended and approved by the relevant committees. Cockburn argues that the main thrust for corporate management and planning came from the officer side of local government rather than from the elected members. According to Knowles (1977, p. 99), many elected representatives are disillusioned with it, feeling that it represents bureaucracy gone mad, and Cockburn (1977, Ch. 1) argues that the position of Chief Officers and committee chairmen has been very much enhanced with the adoption of corporate management, especially in relation to the back-bench councillors. Hence it makes local government appear less democratic than before, and diminished in legitimacy as a consequence.

The last ten to fifteen years have seen an expansion in government employment at both central and local levels, including an increase in the number of jobs available for planners. During this period there has been some reorganisation of administrative structures designed to make them more efficient, comprehensive and better co-ordinated. A third strand, which I touched on in Chapter 1 concerns the relationship between administrators and politicians, for it is within this administrative context that professional scientific rationality comes up against political decision-making. This brings me to a consideration of the role of the state in contemporary Britain, drawing on the theoretical work presented above in Chapter 2.[8] I consider this discussion to be an extension of that offered above, and consequently I shall not restate in detail all the arguments made there, but refer the reader back to them where necessary. I focus on the role of government activity in general, and within that context the activity of land-use planning.

### The Role of the State

I have gone into some detail on the issue of the organisation and management of government departments because this material lends weight to the view of those authors discussed in Chapter 2, who are at pains to stress the power of public bureaucracy. Besides the size and complexity of government administration, they also point to the inaccessibility and impermeability of bureaucracy. This may be partly a result of increased size and complexity, but it is also seen as a result of the quest for scientific status of professional employees, the technical language of some government reports, such as planning documents, and the adoption of mathematical and other techniques, mentioned above in connection with the development of planning as a professional activity.

Conventional constitutional theory involves the assumption that influence flows from the electorate to politicians, either local councillors or MPs, and thence to administrators, local officers and civil servants, responsible to their political masters. There is an assumption that the state is a neutral institution, attuned to the 'general interest' with civil servants and local government officers performing a politically neutral role. This view is implicit in conventional and professional planners' conceptions of the role of public authorities as either arbiters of conflicting pressures with no line of their own or others' making, or as guardians of a 'public interest' somehow in sympathy with the needs of the powerless against the powerful.

In pluralist theory there are two possible roles for government, as distinguished by Playford (1971). On the 'balance of power' view, government accommodates itself to a number of conflicting interests among which a rough balance is maintained. On the 'referee' view, government supervises and regulates the competition of interests so that no one interest group will become predominant. The core function of government is to achieve consensus, and social and political stability. Government itself is a weak institution with no specific interests or well developed position, only to faithfully reflect the various competing pressures which impinge on it from different sections of society. It is seen as an empty vessel into which proposals, ideas, protests and suchlike are poured. Pluralists assume that public bureaucracies are ultimately responsive to a clientele, a controlling agency, or the democratic political process. Planners working within the context of public bureaucracies are answerable to the client public, and land-use planning decisions reflect the balance of interests in a locality, as expressed through political activity. Lack of political involvement is taken to mean satisfaction with the situation. Thus, on this view, all those tower-block residents and commuters who do not organise themselves to complain about their situation would be assumed to be satisfied with it.

This approach does not distinguish between politicians and administrators. Dahl (1961) and Polsby (1963), for example, in their studies of political power in New Haven, pay no attention to the position of administrators. The 'professionals' in their parlance are professional politicians, skilled in political tactics and manoeuvring, and exercising shrewd judgement. They are served by an administrative staff who apparently supply them with the relevant factual information for arguing the points of view they want to promote, the helpful technicians of constitutional theory. This may be the case in the American political system, where top administrative posts are filled by people nominated by the politicians they work with, and who lose office on a change of political dominance. The British situation is different, however, for all civil servants and local government officers hold permanent positions, irrespective of the political opinions of the government or local authority.

The pluralist perspective ascribes an easy openness to the democratic process, and complacently assures us that this is so. As I argued in Chapter 2, several factors call pluralist theory into question in practice. A key point concerns representativeness. MPs or local councillors may quite genuinely lack information of their constituents' views on some issues, or they may be unable to reflect them, due to the constraints of party discipline and the need to follow the party line, especially where majorities are small. Thus, also working against openness are the dictates of the smooth running of the party machine, which in turn facilitates the smooth running of the council or parliamentary decision-making process. An efficient party machine can pressurise would-be rebels into voting a particular way or abstaining, so that decisions may be reached conveniently, with a minimum of debate and delay, and perhaps with a minimum of public awareness. Direct involvement of the mass of the electorate only takes place very occasionally, at election time, where party leaders and manifestos are counter-posed. But national election campaigns are often fought over relatively uncontentious issues, such as 'law and order', which are not necessarily a good indication of policies likely to be implemented once a party is successful at the polls. Local elections also turn upon the popularity or otherwise of the two main parties in the national arena, though in some cases local issues are also important. The crucial role of administrators at both national and local levels must also be noted, for these constitute an army of bureaucrats holding salaried, specialist posts which do not alternate with the swings of the electoral pendulum, chosen for their education and training, and having the 'right' qualifications for the job. Thus they are employed for their professional expertise, not for their political beliefs. It is interesting that planners, except perhaps very senior officers, are not usually questioned on their political views when interviewed for their jobs. Traditionally Labour-dominated councils do not appear to find membership of the Labour Party or commitment to Labour principles a necessary qualification for working in their planning departments. Yet it is these officials who are given responsibility for running the day-to-day routine of government administration, seeking out relevant information and making suggestions and recommendations to be discussed, and often approved by politicians, theoretically their masters.

This is just the perspective of those who stress the power of government bureaucracy, and their whole focus of attention and findings in themselves constitute a strong critique of the pluralist conceptualisation of government. Dennis (1972), for example, is concerned with the impermeability of the planning department involved in decisions about slum clearance in Millfield, Sunderland. He traces a concern in British public administration with the balance to be achieved between administrative efficiency and democracy, where the 1958 Tribunals and Inquiries Act makes provision for the protection of people who feel themselves aggrieved by administrative decisions (Ch. 1). Planning has remained relatively immune, without 'independent, impartial scrutiny and control' (p. 232), having no specific tribunal to hear complaints, and hence little opportunity for redress for those who believe themselves to be on the receiving end of administrative incompetence. Dennis sees the development of public participation in planning as a way of providing a channel of communication between planners and planned, which partly meets this lack. His stress throughout is on bad administration. The Sunderland planners had based their decisions to demolish houses in Millfield either on the 'results' of surveys they had not carried out properly, or principles of obsolescence the Millfield Residents' Association considered to be irrelevant. Yet the planners never justified their position by referring to instructions from the elected members:

> In eighteen volumes of field notes there is not a single expression by a planner of the perfectly sound case . . . that they were the servants of their elected masters, and had to do their best to bring to fruition whatever their masters insisted upon. On the contrary they always insisted in public that what was proposed . . . was on inviolate technical planning grounds and on the basis of impeccable factual data (p. 238).

Dennis's view of government is only hinted at. Reading between the lines one gains the impression that government is well intentioned in principle, but because its structure is complex, and its administration dominated by unelected, unaccountable bureaucrats, there is always the possibility in practice of inefficiency, mistakes being made and a lack of communication and contact with the residents whom the local authority is supposed to serve. He is not attributing maliciousness to local government officers, however. 'Nothing is further removed from the argument and nothing less necessary to it than the postulate of ill-will' (p. 279). It is an issue concerning the institutional framework within which planners work, and are educated, which I outlined in earlier sections of this chapter. In the Millfield situation he sees the planners, not as people but 'as role players caught in a trap they simply did not comprehend' (p. 244). The politicians appear as ciphers, easily manipulated by experts. Fundamentally he is arguing for greater democracy, for an active role for councillors, MPs and grass-roots organisations. Implicit in this view, as also with Weber, is a pessimism about the influence of bureaucrats and their relative immunity to local residents' attempts to put across their requirements and aspirations effectively. Hence there is the need for the safeguard of administrative justice:

> Democracy does not mean that an active citizenry will always involve itself in community affairs. But put at its lowest, and this is democracy's fundamental and ineradicable justification, when it does so the public has a weapon, a frail and insubstantial weapon, but a weapon nevertheless, with which it can face the administration (p. 281).

On this view planners are not sufficiently accountable to their political masters, nor to their client public. Decisions taken reflect the views and values of the professionals, based on some arbitrary standard of fitness, appropriateness and so forth. Here planning could be said to serve the interests of professional planners primarily, whose basic allegiance is to a body of 'scientific' knowledge and code of practice at variance with the requirements of the democratic process. Their position within the bureaucracy gives them the opportunity to impose their own definitions and judgement on the area and people to be planned. If planners are unaware that residents' definitions of needs and standards differ from their own this is hardly surprising, for they are remote and generally inaccessible to the general public, or else do not consider residents' opinions relevant or valid.

Dennis lays great stress on the potential for the mystification of 'scientific' planning procedures, of which he is so scathing in the context of Millfield. I have noted above the development, adaptation and adoption of business management techniques, with the public sector following the pattern of private organisations as part of a move to make local government more comprehensive and co-ordinated, and within it the practice of land-use planning ostensibly less arbitrary, more rational, with explicit processes of decision-making. An explanation of this trend which might be offered concerns the expansion of modern organisations in industrial societies, such that as organisations become very much larger there is a need for conscious techniques of co-ordination and management which are not necessary in smaller organisations, where a few people keep a lot of information, ideas and other considerations in their heads and meet informally in the course of their day-to-day routine. The growth of organisations has occurred in both the public and private sectors, and this expansion might be viewed as a 'modern' phenomenon, 'demanded' by the size and increasing complexity of modern society, and an increasingly specialised division of labour. Cockburn (1977, Ch. 2), for example, offers a different explanation, arguing that new styles of management are sought when existing means of control and organisation are under pressure. This may be due to the rapid growth of a successful firm; the realisation that the firm is failing to hold its own against competitors; or where there is falling profitability. She argues that the adoption of corporate management and corporate planning by the state and the growth of state expenditure have occurred as a result of the acknowledgement of continued persistent problems of poverty and deprivation, particularly in inner-city areas. In their turn, such problems are attributable to 'two related trends: class struggle and the development of capital' (p. 62). People are demanding services. The capital accumulation process requires state expenditure to continue. Earlier methods were recognised to have failed or become inadequate, hence the impetus to change administrative structures and methods of service provision. The argument partly parallels that of Friedman (1977) over class struggle in the labour process, where worker resistance to exploitation and the development of appropriate management strategies are seen as part of the day-to-day reality of the relations of production.

Another writer who emphasised the power inherent in public bureaucracies is Pahl, who drew attention to the importance of urban managers, in its more restricted definition taken to mean local government officers of the various departments. He argued that they play a mediating role between the state and the private sector, and between government and the local population. This view implies that the managers somehow stand outside the state, and are not directly involved in it themselves, a formulation which has been criticised by Norman (1975), for example. Pahl's view of exactly what or who does constitute the state if not, at least in part, its permanent employees, is hazy. In his essay on 'collective consumption' and the state (Pahl, 1977), he does not venture a definition but leaves the reader to assume what is meant. I referred above to Pahl's espousal of a corporatist position, where the state is said to be setting targets for the private sector and directing its activities, albeit only recently:

> In general it could certainly be argued until fairly recently that the state was subordinating its intervention to the interests of private capital. However, there comes a point when the continuing and expanding role of the state reaches a level where its power to control investment, knowledge and the allocation of services and facilities gives it an autonomy which enables it to pass beyond its previous subservient and facilitative role. The state manages everyday life less for the support of private capital and more for the independent purposes of the state (Pahl, 1977, p. 161).

There is no convincing analysis of where these 'independent purposes' emerge from, nor the mechanisms by which private capital loses its dominance. Details of how the state allocates the surplus are not provided, though tantalisingly, the issue is raised only to be dropped, unanswered.[9] We gain no insight into the interests in the name of which the state is said to dominate and command. Further, Pahl's insistence on the usefulness of comparative studies of the role of the state in the Soviet Union and Eastern Europe seems to me to be beside the point, if one is concerned to understand the role of the state in contemporary Britain.

However, if Pahl's view is unclear and vague, the same cannot be said for those writers and activists I called reformist, whose position is explicitly stated. They see the state as a fundamentally benign institution, involved in the conflicting tasks of helping to maintain the viability of the economy and to ameliorate some of its effects, especially extreme personal hardship. It is the only institution thought capable of resolving policy dilemmas and contradictions. They are in favour of some redistribution of resources, welfare benefits, positive discrimination programmes and other reforms. Recent examples in Britain include schemes for the requisitioning and use of otherwise unoccupied housing (Bailey, 1977), abolishing private education and private health care, at least within the context of the National Health Service (Castle, 1976), and securing 'planning gain', some additional benefit for the community as part of commercial development, as described by Jowell (1977). On this view, planners are thought to be accountable to both public and politicians, and specifically to the poor and deprived, even though such people may not be vocal through the formal political process. The assumption is that planning serves business interests, though there may also be some limited opportunity for discriminating in favour of the disadvantaged. The decisions taken thus reflect the basic inequalities of political power and influence in society.

What I have called reformism has much in common with Fabian socialism, exemplified by the writings of Crosland (1956, 1962), Tawney (1961) and Titmuss (1968), and historically by Sidney Webb, who was particularly impressed with the growth and extent of state intervention towards the end of the nineteenth century. He wrote that on 'every side he [the capitalist] is being registered, inspected, controlled, and eventually superseded by the community; and in the meantime he is compelled to cede for public purposes an ever-increasing share of his rent and interest' (1889, p. 46). Frankel comments that Webb

> rhapsodised over this, which later came to be known as 'gas-and-water-socialism'; and was absolutely sure that in this way the capitalists would be eliminated peacefully, and socialism brought in without a hair of anyone's head being put out of place (1970, p. 24).

Fabians believe in the benevolence of the state and the adoption of purposeful government action to supplement and ultimately supplant the market system, though Crosland defends the revisionism of the British Labour Party in abandoning the requirement for outright public ownership, now to be used only selectively and pragmatically to achieve specific objectives. They are enthusiastic supporters of welfare state measures and believe in gradualism, leading to the reform and transformation of capitalism, and ultimately the achievement of socialism through peaceful means.

The contradictions inherent in a reformist position are well illustrated by Donnison's proposals for local service centres, referred to above. Discussing the rash of positive discrimination programmes and community action aimed at redistributing resources in favour of the disadvantaged he comments:

> If these programmes are intended to enable poor people to catch up with richer people — they must redistribute resources, power and status from rich to poor. New political institutions would not be required for this purpose if those who control the conventional machinery of government were wholeheartedly in favour of it. But because a major and sustained redistribution cannot be achieved *without* using the powers and resources of government — the law, the courts, taxation and the social services — initiatives which are confined to voluntary institutions and never win the support of the bureaucracy cannot succeed. Redistribution therefore demands a combination of bureaucratic and political organisation (1973, p. 390).

Implicit, but never spelled out, is some shadowy economic structure somewhere in the background. This is acknowledged in several hints. Thus Donnison remarks that in Britain 'those who do best in the economic market place will generally come out on top in a political free-for-all too' (p. 394) and that 'community action groups which demand from public authorities a response that no one is prepared or empowered to give' (p. 395) will experience frustration and a loss of support. One of his specific suggestions is that people should be encouraged to provide their own services, 'particularly services which the bureaucracy finds it hard to provide: . . . legal advice about racial discrimination, social security rights or the mutual obligations of landlord and tenant' (p. 398), though there is no explanation of why such services are difficult for the state to supply.[10]

As with Dennis, there is a call for 'livelier indigenous politics' (p. 399) in poor areas, as a spur to politicians and bureaucrats; and ultimately recourse to 'the mediating influence of national government, without which the oppressed and deprived seldom make lasting progress' (p. 402). Donnison insists that negotiation between authorities and local organisations in deprived areas is meaningful because everyone can gain something from it. But that 'something' is defined in terms of the *status quo*. This perspective only partly appreciates the potentially conflictual and confrontational nature of protest, especially over housing issues, squatting and race relations, which often involve repressive measures on the part of the state. The possibility that the protesters will find slight reforms totally unacceptable, derisory and insulting does not seem to be considered. Hence this formulation allows him to see class conflict as ultimately benign. Implicitly, the function of the state on this view is to maintain the dominance of existing elites, while perhaps making some small changes in resource distribution at the margins.

This leads me to consider the role of the state for Marxist theorists. In the second chapter I discussed several detailed criticisms of Marxist views of urbanism and urban planning, but I did not look at criticisms levelled at the conceptualisation of the state employed by Castells and others. It is appropriate to discuss this here, to complete this appraisal. Classical Marxism as exemplified by Marx (*The Civil War in France*) and Lenin (*The State and Revolution*) took a simple view of the nature and functions of the state, defined as the executive arm of the bourgeoisie, and essentially antagonistic to the interests of the working class. However, in Britain, for example, the state has clearly provided real benefits for working people, especially in housing, education and health care. Marxists argue that the welfare state can only be understood in the context of class struggle, a concession made by the owners of property and forced out of them by working-class pressure, actual or potential — the ransom the ruling class pays for its survival. How much will be conceded depends on circumstances: the unity and strength of the working class, the state of the economy and the nature of workers' demands. The welfare state may modify, but cannot solve, major social problems, for these are rooted in the class structure of society. George and Wilding (1976) express this position thus:

> The state apparatus is . . . not a neutral umpire, arbitrating impartially between competing groups.
>
> The state will, of course, claim neutrality and impartiality. But this is a false claim in a society where one class owns most of the nation's wealth as well as the means of production.
>
> Social legislation is a peripheral activity of the state; its essential purpose is to protect the system of class-relations prevailing at any one time (p. 91).

Provision of council housing, state education and so forth also have a function at the political and ideological levels in terms of forming ideas and attitudes and maintaining stability, for example by reinforcing the importance of the family through local authority housing allocation policies, housing management and the availability of mortgages. Nowadays there is the promotion of owner-occupation as the 'normal' form of housing tenure by both political parties, relegating council housing to a second-best, remedial status. The growth of owner-occupation has the effect of tying people to stable employment as far as possible, with secure incomes necessary for mortgage repayments. This form of housing tenure stresses the value of individualism and encourages people to become security-minded and property-conscious. Similarly, the school system, besides offering some basic education and perhaps skills useful for gaining employment, also helps to inculcate dominant values, a tolerance for discipline and acceptance of and respect for authority. Further, doctors' exclusive knowledge also has as a corollary a degree of ignorance about health matters on the part of the general public, and a dependence on medical professionals as a result.[11]

This issue of the role of the state has been a subject of controversy between various Marxist writers themselves (Clarke, 1977; Holloway and Picciotto, 1978; Lojkine, 1976; Poulantzas, 1969, 1973; Poulantzas and Miliband, 1969; Mingione, 1977; Pickvance, 1977), though here I am more concerned to look at differences separating Marxist and non-Marxist theorists. The core function of the state is the reproduction and management of existing class relationships, through the formal institutions of the courts, army and police, and informal institutions of socialisation, especially through schooling and within the family. Here planners are seen as state agents, and planning decisions as serving capitalist interests. Basic infrastructure such as water supply, drainage, roads, public transport and so on are provided alongside commercial development undertaken by private firms to enable such developments to function. Buildings are a commodity like any other, and their production contributes to the capital accumulation process. If working people are sufficiently well organised and militant over a planning issue they may win some concessions from the authorities, such as provision of publicly owned housing, parks and open space, or the rehabilitation of existing housing, rather than clearance and renewal. Decisions taken will be a reflection of the strength of the opposing forces in class struggle. The interest of Castells and others is on the potential for radical change in cities, hence their focus on local political activity which has this objective. One of the specific examples Castells uses (1977, Ch. 14) concerns attempts by residents of a Paris suburb, Cité du Peuple, to organise against a proposed urban renewal project. They were largely unsuccessful in contesting the overall scheme, and only won rehousing for a small minority of the households to be displaced. The conclusion Castells draws from this is that these people did not carry enough political weight to force the Paris authorities to change their plans, or to provide adequate alternative housing for them.

Pahl, for example, finds this formulation of the role of the state over-simplified and tautological: if nothing is forthcoming, that is because the state will make no concessions; if there are concessions, that is because they have been won. Against this he suggests that 'it would . . . be reasonable to attribute some measure of credit for the level of public provision to the forces of bourgeois, liberal, humanitarian reformism in the Fabian tradition for ameliorating the harsh logic of capitalist enterprise' (1975, p. 282). Similarly, Pickvance (1976, pp. 201-3) is critical of what he considers to be Castells's undue emphasis on the actions of the movement organisations at the expense of the actions of the local authority. He suggests that this is partly due to problems of access to politicians and officials for researchers, and partly due to a Marxist conceptualisation of state authority, where, by definition, the state is held to be unyielding and uncompromising, only making concessions grudgingly where it feels forced to do so to counteract and soften worker pressure or discontent. Accordingly, this emphasis leads Olives and Castells to overlook the possibility of change emanating from both state institutions and other actors. Like Pahl, Pickvance (1976, pp. 203-7) argues that state institutions cannot be dismissed as sources of (minor) change. The role of the authorities in initiating change is an empirical question and requires an analysis of policy formulation. Pickvance cites case studies of planning and housing issues by Dearlove (1973), Ferris (1972), Lipsky (1970) and Muchnick (1970) as evidence to support his conclusion that popular mobilisation is not the only successful mode of political action, but that success is sometimes achieved by institutional means. Dunleavy (1977, pp. 196-8) criticises the assumption of uni-directionality of influence which Castells and others posit in relations between the protest movement and the authorities or dominant groups, a criticism he levelled at pluralist theorists, as mentioned above. He argues that the assumption that decision-makers will give no concessions unless forced to do so by protest movements is logical where major anti-systemic changes are concerned, but Castells and others go beyond this and say it refers to any change, however small. Thus, this 'assumption of uni-directional influence reduces the study of protest to a very simple stimulus-response model' (p. 197).

Dunleavy (1977b), Pickvance (1976) and Pahl (1975) all argue that the reasons for protest success cannot be located simply in the characteristics of the protest movement itself, and that the actions of dominant groups may have important influences on what structuralists take to be successful protest. Mason (1979, p. 2) has also pointed to the mixed motives of 'progressive' legislation and the political opportunism of Conservative governments. To illustrate this he cites Balfour, in an election campaign speech, reputedly given in Manchester in January 1895 (quoted in Fraser, 1973):

> Social legislation as I conceive it is not merely to be distinguished from Socialist legislation, but it is its most direct opposite and its most effective antidote. Socialism will never get possession of the great body of public opinion among the working class, or any other class if those who wield the collective forces of the community show themselves desirous to ameliorate every legitimate grievance and to put Society upon a proper and more solid basis.

In their discussion of ideology and social welfare, George and Wilding (1976, Ch. 3) give an account of what they call 'reluctant collectivism', exemplified by the views of Beveridge (1943, 1944), Galbraith (1970, 1974) and Keynes (1926, 1936). These writers, they argue, are fundamentally elitist, supporters of inequality, individualism, private enterprise and self-help, though they accept the need for some regulation of capitalism. They acknowledge that the economy is not self-regulating, and that left to itself it leads to unemployment, and more recently inflation. Also capitalism is inefficient and wasteful in its use of resources. However, any intervention in the market must be strictly limited, on this view, which involves an overriding belief in the natural superiority of private enterprise.

Clearly, as Pickvance and Dunleavy point out, there needs to be empirical investigation into the role of central and local government and systematic analysis of policy formulation and implementation. The evidence cited by Pickvance to support his point about local authorities introducing changes is, however, not clear-cut. Two of the studies he refers to do not help to make his point. Both Lipsky (1970) and Muchnick (1970) refer to internal pressure for change from certain sections within the local authority in question, Liverpool and New York City respectively, though this cannot be called 'initiating' change. In both these cases some people inside the local authority were in favour of change, and capitalised on the opportunity presented by protest activity, thus succeeding in getting their minority view accepted within their departments, and changes made. Indeed, Lipsky writes that some officials within New York's housing department made preparations for more radical changes still, in the hopes that the external pressure from local organisations would be sustained and extended, creating a climate of opinion which would allow them to justify yet further changes. Second, Pickvance qualifies his argument by saying, 'admittedly the types of effect concerned can, at most, be described as "reform". A different conclusion might be reached if the stakes were larger' (1976, p. 209). This qualification has the effect of nullifying his criticism on this point. If the stakes were larger success would not be reform, and would not be forthcoming through official channels. This would be true for Castells and Lojkine by definition.

Pickvance and Pahl attribute what they see as local authority willingness to introduce small changes to liberal-mindedness, and there may be an element of this as far as some individuals are concerned. However, at the level of the system, Lojkine (1976, pp. 138-46) argues that one of the roles of the state is to maintain the cohesion of the social formation as a whole, which may involve economic measures, such as investment in firms, regulation of prices and wages, factory legislation, public financing of unprofitable means of communication, and so on. The other role of the state for Lojkine is the maintenance of the domination of the bourgeoisie, involving the facilitation of the capitalist mode of production, the inculcation of dominant values among the population, and if necessary the repression of serious threats against the established order. This formulation points up the fundamental contradiction for the modern capitalist state: that of enabling capitalism to continue to function, while being forced to make concessions to the working class, though not so far-reaching that they would change the basic power relations within society.

I have already noted the growth of state expenditure, which is not normally concerned with direct investment in the production of commodities. However, as I have suggested, the state plays an important indirect role in capital accumulation, taking responsibility for providing services which no individual capitalist is willing to supply, such as health-care facilities, education and training and so forth. Further, the state's involvement in promoting research, providing transport infrastructure and communications networks has the effect of increasing the level of productivity of capital as a whole. The state also serves a vital legitimation function, which helps to stabilise and reproduce the class structure. In this respect, there is a problem of 'diminishing returns' from the provision of services. Once a service is provided, people look upon it as a right, and hence its continued provision adds little in the way of legitimation, whereas a cutback constitutes a source of delegitimation, which will probably be resisted. Because of this, there is constant pressure for the expansion of facilities.

A second area of criticism of Castells and others concerns an emphasis on big business, taken up by Elliott and McCrone (1975) in the context of their study of landlords in Edinburgh.[12]

> As soon as we begin to look closely at landlordism it becomes apparent that much of it remains, as it has always been, in the hands of diverse individuals and families of petit bourgeois character. If we get little indication of this from the Marxists it is hardly surprising for to them the petit bourgeoisie is something of an embarrassment, thought by now to have disappeared, swallowed by the voracious big bourgeoisie (p. 37).

> . . . to the extent that neo-Marxian work stresses the subservience of local authorities to the dictates of big business it will be likely to discount local variation in those provisions which critically affect the lives and chances of urbanites . . . local individuals and institutions *do* have the ability to mediate the directives of central government . . . (p. 36).

According to these authors, municipal authorities are all too often portrayed as the 'lackeys' of big business, and their decisions congruent with the interests of the capitalists and against the working class. Is the role of the local authority limited 'to the creation of the situational advantages . . . on which part of the profits of property capital are based' as Lamarche (1976, pp. 103-4) maintains?

It seems to us highly improbable. Certainly in Britain one can point to many situations where the competition for land involves large and powerful public enterprises, like educational institutions or nationalised industries as well as business corporations and the local authorities themselves are major landowners in their own right. The interests of capital do not invariably win (1975, p. 35).

Elliott and McCrone's contention that Marxists paint a picture of government purely as the servants of big business is echoed neither by Poulantzas nor Miliband, for example. Both these writers acknowledge the existence of various fractions of capital (Miliband, 1969, p. 12; Poulantzas, 1969, p. 72) and the relative autonomy of the state, admittedly a term which is difficult to define. The inter-relationship between capital and state is not the simplistic one suggested by Elliott and McCrone when they claim that the 'interests of capital do not invariably win'. The degree of overlap and inter-penetration between state and capital would seem to be a complex phenomenon, though in the last analysis state intervention tends to be on the terms of capital. Local examples moreover have to be placed within a more general overall framework. While Elliott and McCrone are right to draw attention to the role of small landlords, the authors of the Counter Information Services report are also quite realistic when they stress the limited nature of this role:

> It is incorrect to accept the suggestion made by the government in Fair Deal for Housing that the private sector is dominated by a huge number of small landlords. Most of the landlords, (60%) it is true only owned one house, but these single lets comprise only 14% of all lettings. At the other end of the scale, nearly 40% of all lettings were by landlords with more than 50 dwellings; a further 18% of lettings were by landlords owning between 10 and 50 dwellings (*Labour Research*, 1971, quoted in CIS, 1973, p. 37).

Notwithstanding instances of state regulation of business interests, land-use planning provides notable examples to the contrary, as for example the planning deals over Centre Point and Euston Centre, in London (CIS, 1973, pp. 12-16; Elkin, 1974, Ch. 4; Marriott, 1967, pp. 181-95). Indeed I have been arguing that the scope of land-use planning in Britain is severely constrained by private interests.

Broadbent (1977, p. 207) points out that a formulation which views state activity as more or less directly in the interests of a 'ruling class' may assume 'that the state's economic power is relatively independent of the rest of the economy, so that increased state activity, if required by the "ruling class", can be achieved without threatening the profitability of the private sector'. It is Broadbent's argument that this is not so, and that there is a limit to the amount of state intervention which can take place before private profitability suffers. This is a view that many Marxist writers also endorse (Wright, 1978, p. 154) in their appreciation of state expenditure as ultimately unproductive. This argument has two strands to it. First, that state revenues — taxes — come from a pool of surplus value, and hence there is less left over for accumulation; second, that state spending is not normally concerned with direct investment in the production of commodities. This is an aspect of the inherent contradiction of capitalism: that there is a need for state intervention, but that this must not damage the motor principle of the system — private profit.

A third criticism concerns the emphasis on the movement organisations which Pickvance claims has led Castells to ignore the part played by other actors in the urban development process. In addition to the local authorities initiating (small) changes there is also the part played by landowners, financial institutions and so forth, though he gives no indication as to what this role might be. One possibility would involve developers and landowners co-operating with the local authority in partnership schemes or planning gains agreements. This would undoubtedly be at the instigation of the local authority, though without undermining the commercial viability of development projects, and cannot be thought of as an innovative role. I do not know of instances where financial institutions have initiated changes in favour of working-class interests, where these clash with their own.

Finally, there is Dunleavy's argument that state action may be an important influence on protest failure, through repression, disorganisation and deflecting opposition. This must surely be familiar to Marxists, in theory and practice. When considering the reaction of the local authority to an organisation's activities, Olives uses the concept of repression. The kinds of action that he mentions in this connection include evictions, damage to buildings, harassment of tenants, rent increases, disconnection of services, police support to landlords and so on. Such activity is not necessary where the action taken by organisations is less direct, but the concept retains its validity, though the measures taken — integrating, ignoring and discounting — are more subtle and less dramatic. Examples include the authorities refusing to release information, refusing to meet representatives of the protesting organisation, losing letters or failing to reply, generating confusion and misunderstanding, and making promises which they fail to keep. Further, authorities may attempt to intimidate and discredit individual protesters, or the 'representativeness' of an organisation as a whole, dismissing it as a vociferous minority which does not need to be taken seriously. Another tactic is to identify protest leaders, and in a housing campaign, for example, to rehouse them first, with the likely consequence (and presumably the hope) that the protest will collapse for lack of leadership. Where appropriate, authorities may threaten to prosecute individuals or organisations for some infringement of regulations, or may cut off funding or other resources they have provided for the organisation, on some pretext. They may also dismiss militant tactics of direct action as unreasonable, thus making it impossible for them to have dealings with 'irresponsible elements' who are prepared to take the law into their own hands. In this connection the 1977 Criminal Law Act makes provision for arresting anyone found on other people's property when they have been asked to leave. It affects picketing, occupations of factories, hospitals, town halls, council offices, squatting and so on, and brings trespass within the scope of the criminal law in England for the first time, rather than being treated as a civil offence, as it was formerly.

Needless to say, this is all a far cry from the pluralist view of open government, with a weak polity simply responsive to pressure from below.

The various theoretical approaches considered here all offer different accounts and explanations of the role of government, the interests which it purports to reflect, and within this context, the interests which are served by the land-use planning system. The pluralist perspective has the least credibility, especially in its view of government decisions as reflecting the balance of interests within a community and in its inability to separate out politics and administration. The bureaucratic approach points up serious limitations to representative democracy in Britain, in the stress laid on the influence of government officials. The growth of decision-making techniques is attributed to professionalism, and the expansion of large-scale organisations in this view. The worry is that supposedly democratic institutions are being distorted inadvertently, as a result of the powerful positions of professionals employed in an administrative capacity, and whose alleged technical expertise tends to make them immune from being challenged by laymen — politicians and public. A reformist approach sees the state as essentially benevolent, but unfortunately only able to institute relatively minor reforms, owing to the need to keep the capitalist system going. Great faith is put in the long-term cumulative impact of such reforms, however, and it is assumed that one day capitalism will ultimately be transformed as a consequence. A Marxist view sees the state as basically repressive of working people and their interests, involving a contradiction between maintaining the capitalist mode of production, and being forced to make concessions in the face of worker pressure. On this view the growth in corporate management and other techniques is seen as part of streamlining the bureaucratic machine, basically in the interests of more effective control.

Attributing importance to state intervention and the interests this is thought to serve naturally raises questions about the individuals who work in this context. This chapter has introduced this theme in a preliminary way in the discussion of the development of planning as a technical subject, and the professional view of planning as rational problem-solving. This leads me to a consideration of the possible roles open to professional planners as public employees, which is the concern of the next chapter.

## Notes

1. Hence it is worth pointing out a dissenting opinion. According to Lindblom (1959), evaluation and empirical analysis are intertwined. One chooses among values and policies at the same time, selecting a policy to attain certain objectives and also choosing the objectives themselves. This is to some extent borne out in Levin's (1976) study of decisions concerning the Leyland-Chorley New Town site, and the Swindon Town Expansion scheme, where he describes decision-making in terms of increasing commitment while investigating the feasibility of specific proposals.

2. This not only applies to planning. The same process has occurred in the development of sociology, geography and economics, with the stress on quantitative methodology as part of sociological research design, the adoption of geographical models and the emergence of econometrics. This may be seen as part of the development of relatively 'young' subjects, and also in the context of academic empire-building and the need for relatively new academic departments in universities to justify themselves in their attempts to establish their credibility and to compete for resources with the older established departments.

3. The 'limits to growth' debate illustrates well the complex intertwining of scientific method and political opinion. See, for example, Allaby, 1978, Beckerman, 1974; Kahn *et al.*, 1978; and Meadows *et al.*, 1974. On the point about armament manufacture, it is important to note the work of the Lucas Aerospace Combine Shop Stewards Committee's *Corporate Plan* (1976), discussed by Elliott (1977), which

offers alternative proposals for plant currently producing weaponry. They argue that existing equipment and skills could easily be employed to provide more socially useful products, such as vehicles for handicapped people. So far, management is not interested, however, even though the state would be the likely purchaser of their products, as it is at present, of course.

4. Willer (1971) distinguishes four knowledge systems — magical, religious, mystical and scientific — and classifies them according to their main mode(s) of thinking. Magical knowledge is entirely empirical, depending upon repeated observations; mystical knowledge abstracts observations from the empirical world on to a theoretical level; religious knowledge has a theoretical basis, abstracted on to empirical observations; scientific knowledge combines all three modes of thinking. Despite the fact that science is said to be all-pervasive in Western societies, Willer argues that the prevailing system of knowledge is a modern form of magic, as evidenced by the popularity and dominance of empiricist philosophy (positivism and existentialism) and the continued use of trial-and-error methods of investigation in political and economic enterprise and the social sciences. Modern magic differs from primitive magic in the far greater scope of its effects, which are more conscious and systematic. Magic is an inefficient means of concentrating knowledge since the difficulty of handling increases proportionately as its scope increases. This leads to a division of knowledge in practice, with disconnected specialisms. Feyerabend (1975, 1978) offers a polemic against the rationality of science. Views of the environment not common in the West include religious approaches to nature, as found for example in Hinduism and Buddhism, and described by Allaby (1978), for instance. An interest in other forms of knowledge, such as astrology and intuitive spiritual understanding, has developed among minorities in the West, especially in the late 1960s, as a corollary of the disillusionment some people feel with the bankruptcy of Western rationalism. See, for example, Roszak (1971, 1972).

5. This material is taken from a Southwark Council document on the response to the consultation programme over the North Southwark Local Plan, document number PLA.32/76-77, presented to a special meeting of the Planning and Development Committee on 27 October 1976.

6. See, for example, Buckley (1968), especially papers by Boulding and Bertalanffy, for accounts of general systems theory and some of its applications.

7. This was not a trend he welcomed, however; very much the opposite. 'This passion for bureaucracy . . . is enough to drive one to despair.' The stress on rationality and division of labour reduced people to the status of little cogs in the machine:

> That the world should know no men but these: it is in such an evolution that we are already caught up, and the great question is therefore not how we can promote and hasten it, but what can we oppose to this machinery in order to keep a portion of mankind free from this parcelling out of the soul, from the supreme mastery of the bureaucratic way of life (Weber, reprinted in Coser and Rosenberg, 1964, p. 473).

Merton (1949, pp. 151-60) stressed negative aspects of bureaucracy — rigid over-conformity to regulations and depersonalisation of relationships, detrimental to clients. This view gains common currency in the standardised treatment meted out by bureaucratised medical care, housing departments, treatment of unemployed claimants and so on.

8. I am using the terms 'government' and 'state' interchangeably for the purposes of this discussion.

9. See Pahl (1977, p. 165): 'The crucial questions are who controls that surplus and who benefits from its distribution?'

10. This seems particularly odd in the case of social security rights, since social security payments are administered  wholly by state agencies. One needs to see the

social security system as providing a bare subsistence allowance, deliberately set to discourage 'scrounging'. Establishing eligibility involves considerable scrutiny into personal details, typically done in an officious manner. Many payments over and above the basic rates are discretionary, decided by employees of the Department of Health and Social Security, who interpret general policy guidelines, and with the opportunity of appeal to a tribunal. For the state to provide an effective information service would involve an abandonment of its current stigmatising and near-punitive treatment of claimants. This service is left to the Claimants' Unions, however, which have none of the state's resources to support them. See, for example, Rose (1973).

11. Similar views are put forward by Illich (1971, 1976, 1977), for example, on institutionalised schooling and medical practice. Corrigan and Leonard (1978) give a critical view of social work in a capitalist society, including the ideological significance of the way that social problems are defined, usually in terms of individual failure, incompetence and inadequacy (pp. 99-102). Russell (1935, pp. 47-61) describes the way in which housing separates and isolates individuals and families, and Goodman (1972) also stresses that architecture is not ideologically neutral in discussing 'repressive architecture' (Ch. 4). Fletcher (1976) hints at a deradicalisation process involved in the acquiring of domestic property. These are familiar ideas to Marxist writers. Indeed, Althusser (1969), for example, sees the ideological state apparatus as including education, religion, law. political parties, the family, trade unions, communications and cultural elements.

12. These passages are taken from a working paper prepared for the Centre for Environmental Studies conference on urban change and conflict, York, 1975.

# 4  THE ROLE OF THE
PROFESSIONAL PLANNER

This chapter is concerned with the work of the professional planner, and the various roles open to him or her in professional life. Again I shall draw on the four theoretical perspectives for they each offer different conceptualisations of the planner's role, and suggest various possibilities for professional practice. The pluralist approach to the distribution of power in society allows for an advocate role on the part of professionals. This role, as in the judicial system, is to represent and speak for clients, to act as a go-between or translator between officialdom and unrepresented sections of opinion. The perspective which stresses the power of public bureaucracies treats planners as managers of the urban system, while work in the Marxist tradition views public employees, such as professional planners, as state agents. There are two over-lapping strands to my discussion. One is the way in which the different theoretical approaches treat the role of government employees, and the other concerns the roles that practising planners may want to undertake, according to their own political convictions. The variations between the theoretical perspectives again lead to a somewhat uneven treatment. Much of the literature does not focus on planners specifically, but on local government employees generally, or other professionals such as teachers, social workers or health workers. Hence this discussion of the role of professional planners can be placed in a somewhat wider context. However, before going on to consider the insights offered by the various theoretical perspectives in turn, a few remarks about the planning profession are in order.

## The Planning Profession

The great majority of planners in Britain work for the government, particularly local authorities. Mellor (1977, p. 153) has estimated that approximately 80 per cent of practising planners are accounted for in this way.[1] In an earlier study, Marcus (1969, p. 58) found that young planners were roughly equally divided between working on the preparation of development plans (35 per cent of her sample), working in planning research or teaching in planning schools (38 per cent), and working on development control or detailed designs (27 per cent). More senior planners were responsible for administration. They were barely involved in education or research (10 per cent) or the preparation of development plans (8 per cent), and were most involved in development control and design work (60 per cent of her sample). Marcus noted that the majority of planners come from broadly middle-class backgrounds, indicated by their father's occupation and their own education. She suggests (pp. 58-9) that the political orientation of planners as an occupational group is distinctive, basing this conclusion on interview data. Thirty-one per cent of her sample supported the Labour Party, and 29 per cent the Conservatives. A relatively large proportion supported the Liberals (19 per cent).[2]

Planning as practised in Britain has traditionally been the domain of engineers and architects, who were engineers and architects first and planners second. They received little or no formal training or education in planning as such, but picked it up on the job, and were eligible for membership of the professional institute as a result of their practical experience. Such people currently hold senior positions in very many local planning authorities. But over the years, there has been a growth in professionalisation generally,[3] and similarly a growth in the professionalisation of land-use planning, with expansion of planning education at both graduate and undergraduate levels, and the acceptance of graduates from other disciplines — notably geography, and to some extent economics and sociology — as eligible for planning education and practice, and subsequent membership of the Royal Town Planning Institute. Professional views of planning vary, but it is generally thought to involve rational problem-solving from a position of political neutrality, as I suggested in Chapter 3. Planners pride themselves on being able to take a broad, overall view of a situation, embracing economic and social aspects, as well as engineering, design and environmental factors. It is this co-ordinating perspective which is said to be the hallmark of land-use planning, and which separates it from other professions and academic disciplines.

Planners' professional ideology is of interest here. Foley (1960, pp. 216-18) identified three separate ideologies, variously held by different members of the profession and perhaps at different stages of their careers: the reconciliation of 'competing claims for the use of limited land, so as to provide a consistent, balanced and orderly arrangement of land uses'; the provision of 'a good (or better) physical environment . . . essential for the promotion of a healthy and civilized life'; third, 'as part of a broader social programme' to provide 'the basis for a better urban community life'. Aspects of planners' professional heritage which appear to have been particularly influential include the physical determinism and ideal of low-density residential development embodied in plans for Utopian settlements, and the Garden City and New Town movements, as described by Batchelor (1969), Creese (1966) and Peterson (1968). Other ideals noted by Foley include fostering community life and controlling urban growth.

All three ideologies clearly imply an intertwining of matters of technical expertise and matters of value. This issue is an important one, for there are several opinions concerning what it is that planners do and should do. At one extreme it is maintained, for example by Davidoff and Reiner (1962), that planners are merely a technical resource for the community and political leaders. These authors stress professional responsibility, with planners not imposing their ideas and opinions nor setting goals for the public. At the other extreme, Goodman (1972), Kaye and Thompson (1977) and Palmer (1972) claim that planners are the system's 'soft cops'. This 'managerialist' view is echoed, though without a critical thrust, by Eversley (1973), who stresses the development of the planners' role and influence as 'that of the master allocator of the scarcest resources; land and capital and current expenditure on the built environment and the services which are offered to the community' (p. 342). Reynolds's view is close to Eversley's when she claims that participation 'is an essential element in a theory of planning in which the planners' function is to analyse and synthesize the goals and values of the community' (1969, p. 132).

The growth of professionalism depends upon there being a specific body of knowledge, expertise and skill considered to be the exclusive domain of a professional group, and clearly marked off from other spheres of competence, the 'territory' of other professions. Such exclusiveness is jealously guarded, especially by doctors and lawyers, with long periods of education and professional training and explicit codes of professional practice. Eversley (1973) urges that planners should aspire to the status of these 'traditional professions'. He argues for a comprehensive all-commanding role for the planners, who would move 'to the centre of the stage' in local government, and take key roles in the developing corporate management structures. Such people would not be traditional land-use planners, however, whose training, background, experience and perspective do not fit them for this role, but much more broadly educated. As Eversley rather graphically puts it, 'a new animal . . . must one day soon emerge to replace the present dynosaur' (p. 174).

Undoubtedly, planning education has become more formal and comprehensive in recent years, with an expansion of university and polytechnic courses for both undergraduates and post-graduates, and a broadening of the syllabus to include some study of economics, sociology, statistics, ecology and so forth (Cockburn, 1970; Diamond and McLoughlin, 1973). As with any professional organisation, one of the roles of the Royal Town Planning Institute is to control entry into the profession. This is done by setting professional examinations (RTPI, 1971a) and by requiring that prospective members have completed a minimum period of professional work which the Institute deems relevant — either in local or central government planning departments, private planning firms or educational establishments. Some courses offered by universities and polytechnics are recognised by the RTPI as exempting successful students from taking the Institute's own professional examinations. Such courses are often wider in scope and more intellectually demanding than the Institute's own syllabus, which as Broadbent (1977, pp. 220-3), notes, is still relatively narrow. However, the Institute has effective control over recognised courses, for there is always the possibility of recognition being withdrawn if it is not satisfied with their content, academic standards, political perspective or administration.

Eversley's co-ordinating managerial role for planners, referred to above, has not materialised in a general way. There are differences from authority to authority, perhaps depending on length of service, personality, interest and ambition of the various chief officers. Generally speaking, however, it is the Town Clerk's — now often renamed Chief Executive's — department which is pre-eminent, and in control of overall resource planning and budgeting. In contrast to Eversley, Broadbent (1977, p. 221) suggests a professional insecurity on the part of planners, who have a less established body of knowledge and code of practice than, say, the medical profession or engineers. As a result, the planning profession is sensitive to possible poaching on its preserves and quick to take up new theories and procedures to establish its credibility and enhance its status. Broadbent sees the adoption of the systems approach as one such fashion, as well as corporate planning and, more recently, community involvement and social planning.[4] The development of planning methodology and the adoption of mathematical and other decision-making techniques mentioned in Chapter 3 can also be seen in this light.

Perhaps the most common self-conception among planners themselves is that they play a technical, non-political role. On this view, whatever influence they have on land-use planning decisions is based on technical criteria: land and building surveys, an appraisal of various opportunities for development of particular sites, a consideration of the capacity of the existing infrastructure, assessment of the kinds of development likely to appeal to private development companies and so on. They may not have a personal interest in the area under their jurisdiction, or feel strongly one way or another on the merits or otherwise of proposals for development. They do their job as well as they can, on a strictly technical, professional basis. In one sense this may be thought of as a non-political role. In another sense it is not. In ably serving local government bureaucracy one is also serving its interests, and though these are not always clear-cut, this implies supporting and promoting business interests and the interests of landowners — both urban and rural. It also involves adopting official definitions of 'unfit' housing, 'obsolete' buildings, whether or not single people should have any claim on local authority housing, the illegitimacy of squatting, which sites are appropriate for council housing, gypsies, hostels for the homeless, drug rehabilitation centres and so on. Such definitions clearly involve political choices. Thus, working conscientiously in this context, one takes on these definitions and assumptions, though this may be an unconscious process. The rightness or wrongness of such definitions and the possibilities of other alternatives are difficult to challenge. Indeed, as Palmer (1972, p. 44) notes, to question such assumptions may be to raise doubts about one's neutrality.

The theoretical perspectives outlined in previous chapters suggest four main roles for professional planners, which I shall discuss in turn. First, I consider theorists who focus on the power of public bureaucracies, and view planners as managers of the urban system, firmly located within the context of local government practice.

**Planners as Managers**

Davies (1972), Dennis (1972), Pahl (1975, Ch. 11) and Rex and Moore (1967) all stress the managerial role of local government officers, often depicted as insensitive bureaucrats, out of touch with the needs and aspirations of those to be planned for,[5] able to manipulate and out-manoeuvre the elected council members, and to blind them with science. This is not to say that this is done with any ulterior motive, merely that it is a reflection of the relationship which obtains between full-time professional officers and part-time elected members. At a detailed level one is able to observe a manifestation of this at meetings of planning committees which are open to the public. Few committee members are in a position to challenge officers' recommendations on most agenda items because they may not know very much about planning, or have inadequate information. This is usually limited to reports to read, often at a few days' notice, before meetings. Long agendas mean a lot of reading and a short time given to each agenda item. The committee chairman and Chief Planning Officer are in strong positions by virtue of their authority and can effectively control the proceedings — presenting and explaining proposed development schemes, pointing out details on maps and models, answering councillors' questions. Their tone of voice, the interpretation they put on the various proposals being considered, the assumptions made and the number of possibilities said to be feasible and the amount of detail given about proposals all testify to their authoritativeness. Often councillors have insufficient facts to enable them to reach a decision, and may occasionally ask for an agenda item to be deferred and more information provided. Pressure of time means that many items get very scant consideration, however. The chairman is in a strong position to put across his own view, and may be rarely challenged. But of course the observer in the public gallery only sees those details which come to the surface.[6] Behind this committee procedure is the predominance of officials in the day-to-day routine of the Planning Department. Also 'invisible' to the outside observer is the political discussion of proposals by the Labour or Conservative groups, conducted in private, and especially important in London, where boroughs operate on party lines.

For Dennis and Davies, government officials are inaccessible and almost beyond control by councillors and the public, accountable not to the electorate nor to the politicians who employ them, but to a body of professional knowledge and practice. Pahl has put forward a number of views, stressing the influence of local government officers in his 'managerialist thesis'. He saw them to be crucially important in managing the urban system, and hence affecting people's lives and opportunities, as I noted above. Rex and Moore (1967) attribute the plight of Asian lodging-house residents to the failure of Birmingham's Housing Department to solve its housing problem. They also stress inflexible allocation regulations, involving a five-year residence period to qualify for eligibility for council housing, and discrimination (perhaps unconscious) over standards of property for those Asians who fulfilled this requirement. Thus if any immigrants were ultimately offered council tenancies this was invariably in the worst housing available. However, it is one thing to have some discretion over housing allocation, and effectively to take some decisions which should more properly be the prerogative of elected council members. But it is quite another thing to manage the urban system. Mellor (1977, pp. 149-66) discusses the role of the planning profession and suggests that to some extent these sociologists have taken planners too seriously, accepting accounts such as those of Eversley or Reynolds at face value, and relying on the profession's own estimate of its authority. She argues that by looking critically at what planners do most of the day, another picture begins to emerge. A local authority Planning Department has no substantial budget capital, in contrast to Housing and Social Services Departments, for example, and very many planning decisions are beyond its scope. These include major investment decisions, which are either taken by central government according to its economic and social objectives, or else private individuals, institutions and firms. Similarly, decisions of public bodies such as British Rail, the Central Electricity Generating Board and the National Coal Board may be beyond the control of the Planning Departments, though there is provision for consultation between them and local planning authorities. Also pertinent here are the several detailed limitations to development control which I outlined in Chapter 1. Indeed it has been my argument throughout that land-use planning is seriously constrained by the private ownership of land and capital and private initiative in development. In fact at the local authority level, much routine development control revolves around planning applications for extensions to existing dwellings, renewal of short-term permissions for temporary uses, design and layout of housing schemes, maintenance of Conservation Areas and suchlike.

Thus Mellor argues that it is quite out of the question for planners to direct the urban system as too many of the pressures for change originate outside the locality. She suggests that nowadays few sociologists and perhaps few planners would agree that the planner is 'the master allocator of resources' as Eversley claimed. Because of the limitations to planning powers the

> calls for a positively discriminatory approach in the allocation of urban resources have to go unanswered by the planners, for there are only limited possibilities to their shifting this previously determined allocation in favour of one group rather than another without control of the legal rights in land (Mellor, 1977, pp. 162-3).

Planners do have power in local circumstances, particularly in connection with slum clearance, urban renewal, planning zoning and so on. The fault these writers have made is failing to distinguish between what planners do in one specific situation, such as Millfield as described by Dennis, and Rye Hill (Davies, 1972) and the total effect of the planning system as a whole. This view is echoed by Broadbent (1977, p. 173) who stresses the 'gap between plans and powers', since the planning system does not directly implement development. For urban planning in Britain involves a fundamental contradiction: that of undertaking the very real task of regulating the use of land without any real power or resources for implementation. This leads, in practice, to planners 'myopically getting on with the minutiae of day-to-day development control' (Broadbent, 1977, p. 173). Hence the managerialist thesis which claims that planners and other local government officers effectively control the urban system does not hold, and Pahl (1975, Ch. 13), for example, has revised his earlier position. In a more limited way, however, local government officers, including planners, do have considerable influence over people's lives and opportunities.

Indeed, land-use planners have been much maligned in Britain in recent years, and blamed for poor living environments and social problems said to be related to them. They are also associated in people's minds with delays, red tape, rules and regulations, unimaginative and mediocre designs and proposals. As Mellor (1975, p. 278) remarks, it has become commonplace, 'a cliché of the political platform' for the present economic crisis to be re-interpreted as an urban crisis, and in this process planners have come in for considerable criticism. On this view, unemployment, poor housing in inner cities, squatting, racial tension, crime and delinquency, public transport costs, vandalism and truancy are all said to be to do with 'the failure of the city'. It is thus only a short step to holding planners — the city's professional guardians — responsible for its alleged shortcomings.

One of the strands sometimes associated with this formulation concerns the importance of local political action over this ostensible deterioration in the quality of life, expressed through local pressure-group activity. An acknowledgement that people who are poor, unskilled, unemployed, or educationally deprived do not have the same ability as other groups to permeate the political system has led to professionals working for local organisations and individuals to help them to articulate and make their views known through the established political process. This situation provides the opportunity for an advocate role for planners and other professionals, who can speak for 'client' groups in the political arena, analogous to the role of the lawyer who argues a case for his or her client in court.

**The Planner as Advocate**

An advocate role involves representing people who might otherwise not be able to put their points of view through the formal channels of communication provided by the political process, and several writers have urged that this role be taken more frequently by professionals (Dennis, 1972, p. 246; Hunter, 1969; Pinker, 1968). Some professionals work voluntarily in this capacity, or are paid by independent institutions such as charities or bodies which sponsor research, though some are paid by government departments. The following examples illustrate this.

A neighbourhood action project in an inner area of Liverpool, sponsored by Shelter, the national pressure group for housing and homelessness, employed community workers to organise people to press for better housing conditions, and they acted as advocates of residents' interests in dealings with government departments (McConaghy, 1971, 1972; Rose and Puckett, 1973). Professional community workers, whose focus of attention is at the level of the community rather than the individual case-work orientation of traditional social work practice, from which community work has developed,[7] also help their 'client' community to articulate needs and preferences within the political system, while being employed by local government. People who worked in the Community Development Projects in different parts of the country in the late 1960s and early 1970s also took on this advocate stance in organising people in various campaigns for better housing conditions, job opportunities and so on, and in publishing the results of these experiences. The Home Office provided salaries and funds to cover office premises and administration. It continues to give Urban Aid grants for specific local projects, perhaps concerned with providing legal advice, housing advice, a community planning centre or adventure playground. Applications for Home Office funding through the Urban Aid programme must be supported by the recommendation of the appropriate local authority, which provides 25 per cent of the budget. Some planners and architects are also involved in working with local organisations in Britain. They may be hired by, or offer their professional services to, community groups to provide specialist expertise — architectural drawings, detailed layouts, knowledge of planning law and procedure, helping to prepare and present the case that an organisation wants to argue, perhaps acting as expert witnesses at public inquiries in support of alternative planning proposals put forward in opposition to those of a local authority or development company. Students and staff of architecture and planning schools may act as a resource to local organisations which do not have the necessary technical knowledge amongst their members, though in practice this does not appear to be as common as it might be. Working within local Planning Departments or the Department of the Environment, there is the possibility of advocacy of a different kind. Instead of being in direct contact with a specific group and undertaking to represent their interests, the planner can support the (assumed) interests of particular sections of the community in discussing routine planning proposals, recommending, for example, that the needs of the elderly, the handicapped or mothers with small children be taken into account. Albrow (1970) describes how sociologists may be employed in an advocate capacity, representing the opinions of particular sections of the 'client population', and paid by the local authority. Much advocacy is of course unpaid, with professionals contributing skills, or being regularly involved in community organisations in their own time, working on campaigns concerning local housing and planning issues, for example, offering a legal advice service, advising on tenants' and welfare rights, or writing for community newspapers.

The advocate role has a place in both the pluralist and reformist approaches to the distribution of power in society, and my treatment of it here needs to be placed in this context.

Associated with a pluralist view, advocacy may be thought of as a reformist posture. Mazziotti (1974, p. 44) argues that advocacy planning in the United States has had as its central assumption 'the uncritical acceptance of a social myth called pluralism'. Advocates are employed to represent the interests of those people who are either unrepresented in the official political system, or insufficiently articulate for some reason. The assumption is that if only people are given the opportunity to be heard, then their opinions will influence the outcome of government decisions. Davidoff (1965, also reprinted in Faludi, 1973) and Peattie (1968) are enthusiastic proponents of advocacy in the American context. Davidoff argues that it is advantageous for the planning process on three grounds: as a means of better informing the public of alternative choices open, and forcing the public agency to compete with other planning groups to win political support. Indeed, he goes so far as to suggest that in the absence of any opposition or alternative plans presented by interest groups, public agencies have little incentive to improve the quality of their work. A third advantage is said to be that it puts the onus on those who are critical of the establishment to produce superior plans. Mazziotti (1974, p. 46) takes issue with this kind of complacency, with its acceptance of the need for gradualist or reform-oriented solutions to social problems, and with its assumption that 'the prime criteria for effectiveness  must be the existing institutional setting'. Indeed many people who have worked as advocates have acknowledged the severe limitations of the pluralist perspective, both in theory and in practice, a disillusioning experience, as accounts by Blair (1973, Ch. 12), Goodman (1972) and McConaghy (1972) all testify. As Goodman puts it, advocacy planning allowed the poor 'to administer their own state of dependency' (p. 212). He summarises his own experience:

> In a highly technical society, we argued, the availability of technical help to all groups was a critical requisite for true power sharing. The use of their own experts in planning and architecture was going to give the poor a strong voice at the places where decisions about their lives were being made. Indeed, we were able to delay or make changes in some urban-renewal and highway plans. But we were to learn the limited extent of our influence (p. 211).

The poor could direct their own welfare programmes, have their own lawyers, their own planners and architects, so long as the economic structure remained intact — so long as the basic distribution of wealth and hence real power, remained constant (p. 212).

The need and opportunity for advocates has been seen as a real extension of the democratic process by its proponents, and as an indictment of the effectiveness of established political institutions. Its existence is anyway a criticism of the pluralist approach, for if competing interest groups were equally placed in the political process, and did indeed carry equal weight, which the notion of 'countervailing power' suggests, there would presumably be no need for advocacy in the first place. The serious weakness of the pluralist perspective in the context of Britain and America has already been considered in Chapter 2, and this discussion need not be repeated here.

Associated with a pluralist view is a 'choice theory' of planning, set out initially by Davidoff and Reiner (1962), an ideal, theoretical approach, where planning is defined as a process for determining appropriate future action through a sequence of choices, with the ultimate objective of widening both individual choice and the efficiency of the urban system. Choices are made at three levels: the selection of ends and criteria, the identification of a set of alternatives consistent with these general prescriptives, selection of the desired alternative and the guidance of action towards determined ends. The stress is on individuals and groups making choices, for example, between different housing types, between a variety of job opportunities, several transport and recreational facilities. Choice is considered to be a good thing in itself, and one of the roles of the planner is to spell out as many varied ways as possible of widening the choice of facilities and amenities available. This emphasis has been influential in land-use planning in Britain, and was somewhat ritualistically built into the initial statement of aims and objectives of many plans in the 1960s, though it has been less prominent with the recognition of more stringent economic conditions in the 1970s. Both Mellor (1975, pp. 121) and Pahl (1975, Chs. 7 and 8) argue the counter-case: that the emphasis is in fact not on choice, but constraint. A choice theory of planning mirrors notions of 'consumer sovereignty', with the environment and social facilities of all kinds viewed merely as commodities to be consumed, and indeed Broadbent (1977, pp. 204-7) calls pluralism a 'market theory'. The assumption is that desired facilities are in adequate supply, and within people's reach financially, which is clearly open to question. 'Choice' is only meaningful provided one can afford more than one type of housing, or if there are several equally convenient neighbourhoods to live in, several suitable jobs to choose between, and so on. Yet this is patently not the case for very many people.

Advocates will usually identify with the values, views and priorities of the people with whom they work. In the sphere of land use planning this often leads to a demand for 'real' participation in the planning process, not merely consultation of a more or less tokenistic nature. The existence of different definitions of participation is important and has led to much confusion and disillusionment in practice, discussed more fully in the following chapter. In this connection it is worth considering the 'representativeness' of the advocates and their ability to reflect faithfully the wishes of their client group without losing or adding anything in the translation. This is one of the dilemmas that concerned Peattie (1968), despite her initial enthusiasm for advocacy, for she recognised a temptation to 'educate' her client organisation with her perception of their situation. This very point is commonly taken up by local politicians, who, for their own purposes, are often at pains to discredit the representative nature of articulate, antagonistic organisations and want to dismiss them as a small vocal minority, 'politically motivated young men', professional outsiders with no long-standing links with the area, or a few people determined to wreck things. Undoubtedly, a professional may sometimes want to re-interpret and renegotiate his or her brief, though this is probably a subtle or even unconscious process, especially where an organisation is reliant on such a person for their expert knowledge and not in a position to query advice which may be offered. Professionals who are involved in a general way as community activists rather than hired experts must obviously be very sensitive to this issue, or they may run the risk of losing their 'constituents' if their leadership becomes too directive. The problem of defining interests may easily lead to severe argument in practice, with the 'advocate' isolated from more 'radical' or 'reactionary' opinion from within the 'client' community. A local organisation, or some people within it, may want to oppose commercial redevelopment in their neighbourhood, for example, or the conversion of unused industrial premises to shops, restaurants and bars, because they feel that such uses are not in local people's interests. They may find the ground cut from under their feet if local people disagree, and are willing to take jobs working in restaurants, cleaning offices and so on. In a particularly blighted area, *any* development may be welcome to residents, whereas the advocate professional may urge continued opposition to planning proposals until the 'right' proposal comes along. On this point, Wilcox and Richards (1977) have argued that it is no use planners or local organisations trying to oppose market trends, but that they should attempt to harness these trends to their own objectives as much as possible. These authors criticise many local organisations active in land-use planning in London for negativeness, lack of vision and opposition to development proposals, which, they claim, has resulted in continued blight in many areas.

**Reformist Roles**

A third approach to be considered is that of reformism, where practising planners try to use their training, experience and position to secure some benefit to groups disadvantaged by the market, following the dictum that planning should be compensatory, as argued by Gans (1968). This might involve working out and implementing a variety of administrative proposals, for example for the use of otherwise empty housing (Bailey, 1977), the temporary use of derelict land and buildings, for adventure playgrounds, gardens or community uses (McCulloch, 1978; McKean, 1977), or urging that planning gains be forthcoming before planning permission is granted. In his discussion of planning bargaining and the scope of planning gain agreements, Jowell (1977) notes that planning officers appear to be important in taking the initiative and encouraging their councils to bargain for some additional planning benefit to be provided. Broadly speaking, there have been no public guidelines, with a few exceptions in London.[8] Planning gain negotiations, where planners and developers are involved in a bargaining process, as discussed above in Chapter 1, are opportunistic, and conducted by a small number of senior planning officers in secret. The planners aim to secure some 'community benefit', while the developer hopes to concede as little as possible in order to secure a favourable planning permission. Very many of the bargains made are done by agreement between the development company and the local planning authority, a point of interest to Jowell, as a lawyer, who notes that legal procedures 'exist not only to control the developer, but also to avoid the abuse of official discretion' (p. 433). An editorial in the professional planning journal (*Journal of the Town Planning Institute*,

1972, p. 340) also stressed that planning gain negotiations should be conducted wisely, and that planners should not take advantage of their professional position.

There are other possible roles for the reformist: in academic teaching and research, and local politics, perhaps working for voluntary organisations. The advocate role discussed above in connection with pluralist theory may also be seen as a role for would-be reformers. A dilemma may concern the nature of the employing or funding institution. Teaching, research and community work are usually financed by government in Britain, which also sets the specifications for the job to be done, albeit usually implicitly. Such workers may thus be in the contradictory position of organising opposition against the local authority or government department which is also their employer. Lewis (in Cowley *et al.*, 1977, pp. 121-3) for example, mentions social workers supporting gypsies and squatters against their employing authorities.

The question of the 'research industry' should be raised here. There are many aspects of our society which are not well known, and which suffer systematic distortion and misrepresentation at the hands of the media. Examples of alleged 'facts' which have passed into conventional wisdom include the ostensible disappearance of poverty, tales of subsidised council tenants earning £100 a week, with their expensive cars parked at the door, accounts of squatters who occupy houses while their owners are away on holiday or out shopping, the difficulties of getting rid of unwanted tenants, such that landlords' lives are made a misery, and so forth. In one sense more information is required, and should be more generally made available, whereas there are few outlets for information which challenges the prevailing views disseminated through the established media, and those which exist are not particularly accessible, mainly confined to 'left-wing' or 'alternative' bookshops. Examples of such material are reports published by Counter Information Services, Social Audit, Labour Research Department, newspapers and pamphlets issued by various left-wing political parties, and leaflets written in connection with specific campaigns.[9] Yet very little notice is taken of these publications by government or the established media, perhaps because they are not well known, but fundamentally because they are too critical, for their analysis often calls for or at least implies the need for radical changes in British society.

Government continues to fund expensive research projects and ignores some research findings. A good example of this is provided by the Urban Programme which involved relatively small grants to particular areas, defined as deprived, as for example Educational Priority Areas, General Improvement Areas and so on. There was also the setting up of the Community Development Projects with an explicit research dimension, and a brief which defined deprivation in terms of individual incapacity and inadequacy on the part of deprived people. When CDP workers redefined the whole issue, and attempted to link inner-city poverty with economic questions underlying the entire socio-economic structure of British society, the projects were disbanded and the money went elsewhere. Local authorities continue to dissociate themselves from some of the specific findings and interpretations published in the CDP reports. Since then, however, there has been more small-scale, specific investment in inner-city areas, through the Inner City Programme, and partnership schemes, all as if the CDP work had never been done. The limited parameters of government-sponsored research will be familiar to many who have written research proposals applying for funding. Recently, for example, proposals to investigate transport facilities and connections between the journey to work and unemployment have been sponsored by government, despite the body of work which argues that much unemployment in Britain stems from structural changes in the economy, the adoption of automation and so on, rather than unemployed workers merely finding it difficult to get to work. Government-sponsored work on squatters in London was refused publication by the Department of the Environment, presumably because it was favourable to squatters, presenting them as ordinary people who repaired and improved property, in contradistinction to their public and press-promoted image of scroungers, free-loaders and vandals. This study has subsequently been published by Shelter (Kingham, 1977).

Harvey (1973, Ch. 4) discusses several roles for the academic, and in this context distinguishes between three kinds of theory: *status quo* theories, rooted in reality but preserving vested interests, counter-revolutionary theory, also serving vested interests, but distorting reality in some way, and revolutionary theory, grounded in reality and interpreting it in such a way that points to a change in the *status quo*. He is discussing theories which purport to explain the formation of ghettos, and insists that it is merely counter-revolutionary to engage in 'yet another empirical investigation of the social conditions of the ghettos'; 'mapping even more evidence of man's patent inhumanity to man is counter-revolutionary in the sense that it allows the bleeding heart liberal in us to pretend that we are contributing to a solution when in fact we are not' (p. 144). Nor should we indulge in that 'emotional tourism which attracts us to live and work with the poor "for a while" in the hope that we can really help them improve their lot' (p. 145). What is needed is the self-conscious construction of a new paradigm, and commitment to revolutionary practice. As academics, Harvey writes, 'our task is to mobilise our powers of thought to formulate concepts and categories, theories and arguments, which we can apply to the task of bringing about a humanizing social change' (p. 145).

An extension of the reformist role which might be adopted by people working within local and central government is that of the 'bureaucratic guerilla' (Altshuler, 1965, Ch. 8; *Community Action*, February 1972, pp. 26-8), where officials may provide information that would otherwise not be available to organisations with which they sympathise and which are in conflict with the local authority or government department. Thus they are working closely with an organisation, and have access to inside information which they are prepared to pass on. This possible role may be considered highly unprofessional or unethical, and it is scornfully repudiated by Eversley (1973), for instance. In any case it would have to be done discreetly, or else probably not for long. The contact with the organisation itself may need to be kept quiet, so that it is difficult to trace the source of 'leaks' with certainty. Hence community activists just happen to come across confidential reports, details of planning proposals, lists of empty property, instructions to the local authority from central government, information concerning which councillors and officers are prepared to support particular proposals, and hence be allies in a potential campaign, informal housing allocation policies and so on. This is a conflictual role for a practising planner, and one that perhaps few would want to take up, involving as it does a fundamental contradiction for the individual concerning his or her loyalties, and the possibilities or personal repercussions. Probably the easiest time to pass on such information is if one is leaving a particular local authority, for the likelihood of repercussions is thus minimised. Some local government officers may find this role worth while, and use it partly to justify their form of employment. Being able to help people in conflict with the local authority thus becomes a way of reconciling their ideals with the need to earn their living. This role is not confined to the point of view I have called reformism, and clearly overlaps with Marxist practice, and this leads me to the fourth contribution to be considered — a Marxist conceptualisation of state employees as state agents.

## Planners as State Agents

Structuralist Marxists whose work was mentioned in Chapter 2, such as Althusser, Poulantzas, Lojkine and Castells, all see state employees as implicit agents of the dominant class, operating within the confines of roles which are 'given' by the system, working within a structural strait-jacket. Castells, for example, expresses it thus: agents 'whose most obvious expression is in social classes, are only the supports of these structural relations' (1976, p. 150); individuals are 'actor-supports, those men-who-make-their-history-in-determined-social-conditions'(p. 155). For Castells, urban planning has a repressive, or at best reformist, function, though it is probably true to say that most professional planners would be outraged to be categorised as oppressors.

I suggested above that there are several different positions within the Marxist tradition, and the question of structural determinism is one which marks a cleavage between such writers as Poulantzas and Castells on one hand, and Miliband on the other, together with what Lukes (1974) calls 'humanist' or 'subjectivist' interpretations, of such writers as Sartre and Goldmann, for whom the 'subject' 'has a crucial and ineradicable explanatory role' (Lukes, 1974, p. 52).[10]

This distinction is exemplified in the debate between Miliband (1969) and Poulantzas (1969), where Poulantzas criticises Miliband for not seeing the state as an objective structure, and giving the impression that it can somehow be reduced to interpersonal relationships between members of groups which make up the state apparatus — politicians, civil servants, the judiciary, military, police and so on. In Poulantzas's view the state is a structural category — a repressive, controlling institution, enabling capitalism to continue to function by largely facilitative, interventionist measures. Local government constitutes local outposts of this central institution, and is the arm and manifestation of the state in a local area. The hardline structuralist view sees state employees as operatives running the machine, with a very limited range of appropriate behaviour and responses to particular requests, demands and situations. Miliband counters, claiming that Poulantzas overstates his case, with its stress on structural constraints, which are so compelling 'as to turn those who run the state into the merest functionaries and executants of policies imposed upon them by "the system" ' (1970, p. 57). Poulantzas (1973) defines 'power' in terms of structural determination, whereas Lukes (1974) wants to distinguish an 'exercise of power' which involves two assumptions: that 'it is *in the exerciser's or exercisers' power* to act differently' (p. 55, stress in the original) and that responsibility for the consequences of action can be attributed to particular individuals or groups. Lukes notes the similarity between this distinction and Mills's (1959) differentiation between fate and power. His 'sociological concept of fate' had to do with

> events in history that are beyond the control of any circle or groups of men (1) compact enough to be identifiable (2) powerful enough to decide with consequence and (3) in a position to foresee the consequences and so to be held accountable for historical events (Mills, 1959, p. 21).

Mills argued that an exercise of power can be attributed to people in strategic positions 'who are able to initiate changes that are in the interests of broad segments of society but do not' (Lukes, 1974, p. 56), and that they should be held responsible for the consequences of their actions or inaction. There is an important question, with both theoretical and practical implications, which should be noted here, for we are left with the problem of defining where structural determination or fate ends, and power and responsibility begin.

A less determinist position would be to accept a Marxist view of the state, but still to see the possibility of changes being made, and state employees being in a position to make a contribution to this process if they want to put their energies in that direction. On this view, structures are no more than man-made historical institutions and as such they can be changed if enough people want to do so enough. Corrigan and Leonard (1977, p. 104), in their discussion of social work practice, reject the view that 'state employees . . . are simply henchmen of the ruling class', for this fails to recognise what they call the semi-autonomous nature of the state, and gives too determinist and monolithic a picture of the state apparatus. However, they insist that state employees are not neutral, though they may affirm their neutrality, and point to the reliance on science as the basis for policy decisions in this connection: 'The growth of professions, especially those employed by the State, such as medicine, teaching and social work, depend upon claiming neutrality on the basis of a particular body of objective knowledge'

(p. 104). Moreover, welfare bureaucracies form part of the ideological state apparatus, and as a result have more autonomy than directly repressive sections such as the police and the army. Lewis (1977) has argued that some social workers have undergone a radicalisation process and see their interests as similar to those of their working-class 'clients' rather than the state employer. Similarly, Donnison (1973, p. 391) hints of public officials 'going native'. Lewis argues that social workers and other state employees must be convinced that 'they should use the power they have as workers within the state apparatus to aid the struggle of the working class and to weaken the attacks of the ruling class' (p. 122), and that state employees should organise alongside other sections of the labour movement, through the trade unions and also in specific campaigns. Corrigan and Leonard (1977, pp. 141-57) explore possibilities for a Marxist practice within social work, emphasising 'the possibilities of action' in distinction to the 'brick wall of fatalism' (p. 141). They recommend that social workers form alliances with working-class organisations, that they join a union, and that they recognise the limitations of working-class organisations which have formed in the context of capitalism, and are mainly interested in economic improvements in their position, rather than challenging the fundamentals of the whole system. They suggest that social workers should try as far as possible to work collectively with colleagues and not be arrogant or risk alienating possible support from colleagues and other staff. A recent example of joint action by trade unionists and professionals in London would be the campaign against the proposed closure of the Elizabeth Garrett Anderson Hospital for women, which has been in dispute for some years. Here a vigorous campaign has involved the hospital workers and Health Service administrators in several unions (COHSE, NUPE, NALGO), other interested trade unionists, members of various political parties (Labour Party, Liberals, Communist Party), local tenants and community associations, and members of women's organisations. The outcome of four or more years' campaigning was a reprieve for the hospital in May 1979, when the newly elected Conservative government allocated funds, but specified that its future role should be limited to work in the field of gynaecology rather than providing a full hospital service for women. If state employees are the people in the middle, administering state services to 'clients', there is the possibility that they can form alliances either way, with the state or with their 'clientele'. In the case of the EGA and other hospitals faced with closure, the Health Service employees clearly allied themselves with hospital users against government policy of reducing public expenditure.

Kaye and Thompson (1977, pp. 101-7) point to what they see to be a crucial dilemma faced by state employees, writing specifically about planners, namely: 'how to work towards expanding those areas of state intervention which improve living standards by socialising resources without becoming co-opted in the process' (p. 104), and ask what are political ways of working as a planner. They suggest that the question has been wrongly formulated in Britain, in terms of working 'inside' or 'outside' the system, but this is a somewhat false distinction, as one is always within the system to some degree. In working with community activists one has the opportunity to make available one's professional knowledge and expertise for local organisations, though they warn against the like-lihood of becoming incorporated, as is the case with advocacy planning. The limitation of this role is that it masks the common class position of various groups and plays up their differences, and it ignores the relative powerlessness of such groups against the power of the state. Moreover, the state encourages community-based action, they argue, due to an understanding of the weakness of fragmentation. Resources may be given to the least conflictual groups, involved with conservation, for instance, Improvement Areas or Housing Action Areas.[11] Local organisations have to compete against each other for scarce resources such as grants from the local authority, or information. This can take considerable time and energy, and may tend to lead to competitiveness rather than co-operation between organisations which are fundamentally in the same position, even though those people involved undoubtedly see through this strategy of divide-and-rule.

Kaye and Thompson maintain that there are 'genuine opportu-nities to work in socialist ways' (p. 106) either within the state bureaucracy or with community organisations. They do not go into details of what this might amount to, but stress the need for clearer links between political activity and the work of planners. They also urge that state employees work in a trade union, and have a consciousness of themselves as workers rather than professional intermediaries:

the important starting point for the planner is the consideration of ways to increase both the demands for and the satisfaction of those collective needs which by virtue of their non 'profitability' pose a threat to the stability of capital (p. 107).

Examples of links between planners' work and political activity might include housing campaigns on behalf of tenants, the homeless, squatters, perhaps involving requisitioning, and bringing otherwise unused housing into use; cuts in government spending, involving a decline in local services, such as hospital closures; campaigns centred on employment issues; contacts with anti-Fascist activities, countering racialist explanations of urban problems; women's campaigns for the provision of nursery facilities to allow more mothers to work outside the home, and so on. Kaye and Thompson argue that there is a need for clear articulation by planners of allegedly urban problems as stemming from the social and economic conditions of a capitalist mode of production.

Their perspective is in contrast to that of Goodman (1972) who argues that planners and architects working for the state are bound to implement its interests, acting as the system's 'soft cops', for example, holding tokenistic participation exercises. He concludes that the only thing to do is to get out of state employment and work for community organisations. There appear to be two theoretical objections to this formulation. First, there is an implication that the state's interests are unitary and monolithic, and, second, that one can really be on the outside. Kaye and Thompson make the point that one is never beyond the scope of the established political and economic order, and especially if one is trying to change aspects of it. One has to know about it, to relate to it, and always be alert to the likelihood of repression and incorporation.

Cockburn (1977) argues that working-class people involved in local political campaigns see bureaucrats as state agents, and hence as part of their problem. Protest groups particularly have this posture, and often a disrespectful, confrontational style. Bureaucratic procedures involve spurious classifications of people, as for example with the distinction drawn between families as opposed to single people in local authority housing allocation policy, and the stress on 'the genuinely homeless' as distinct from 'politically motivated' squatters. State employees usually insist on talking to a leader of an organisation rather than a collective. They appear only to understand groups that constitute themselves similarly to bureaucratic structures, with committees, accounts, agreements in writing and so on, hence the apparent need for parallel structures in order to deal with the bureaucratic apparatus. Cockburn discusses these issues in connection with squatting in Lambeth (1977, pp. 76-87), and stresses the social control aspects of 'the community approach' (Ch. 4), involving public participation, tenant participation in managing housing estates, the possible role of Neighbourhood Councils and community development work.

Most, if not all, who accept a Marxist analysis of society also want to see radical change come about, and are faced with the dilemma of how to earn a living and put their energies to what they consider worthwhile uses. Perhaps not many professional planners share this perspective. If they do not, they do not have a personal problem, though they are part of other people's problems — those who are up against the state apparatus. For those who do take this view there is both a pessimistic aspect, and an optimistic one. For the extreme structuralist position there appear to be only two ways out: revolution or cynicism. If small changes are merely reformist tinkering — despite the fact that this may improve conditions in a limited way, or in a limited area — then working as a planner will be a dissatisfying experience. Research and teaching may provide slots for such people, shielding them to some extent from the contra-dictions involved in holding this view of the state and state employees, and being so employed themselves. The hard-line structuralist position has little to offer the would-be reformist or radical employee, apart from a rather 'schizophrenic' existence, working for the system during working hours, and against it in his or her 'free' time. There is a hopelessness and nihilism in this analysis that tends to lead to defeatism and cynicism about bureaucracy, and the individual's role within it. It is this frame of mind that can lead people into thinking that it is an achievement to do nothing. What is the point of trying to bring about changes, if one believes oneself to be doomed to be beaten by the structure? Further, it is very difficult to work at something on a daily basis that one categorically does not believe in, for it involves a complete denial of oneself. A more optimistic approach is shared by Corrigan and Leonard (1977), Kaye and Thompson (1977) and Lewis (1977), for example, though none of these writers suggests that their objectives for everyday profes-sional practice are clearly defined, nor are they naîvely optimistic that it will be easy.

This chapter has been concerned to describe various roles for the professional planner based on different theoretical perspectives and depending on his or her view of what constitutes valid professional practice, and personal motives for working in local or central government. As far as rewards go — salary, status, promotion, responsibility — the various roles offer different possibilities. The conscientious career planner working within government bureaucracy will be rewarded by promotion and increased responsibility in a way that a person who is highly active politically may not, though a touch of reformism will probably not be a hindrance. Working for local groups as an advocate or community planner will be more precarious financially, and perhaps be looked on with some mistrust and suspicion if one later wants to work for government. Private practice concerns only a small minority of planners. It is the most highly paid, and having no responsibility for plan implementation, has no development control function. This work may seem attractive in that it involves research and formulating planning options for a particular area, and is free from administrative and political constraints of local government work. I suggested earlier that private consultants are very much dependent on government for contracts, either in Britain or, more commonly, overseas. Their work is never 'finished', merely handed over, and they do not get the opportunity to follow things through to the implementation stage. A corollary of this 'independent' status is that their reports can always be shelved, and government can completely dissociate itself from their findings and recommendations if it wants to. For those with reformist or radical political aspirations the professional role will provide contradictions arising out of the possibilities and constraints offered by their structural position as state employees, and their political ideals. One of the justifications for working in local government may be to be able to help individuals and organisations with whom one sympathises, and which are in a more conflictual relationship with the authority, for instance tenants' associations or community action groups. Such organisations, together with conservationists, squatters, businessmen and interested individuals, may be actively engaged in trying to influence land-use planning policy and practice, and I shall turn now to consider the scope for involving the public in planning decision-making in the next chapter.

## Notes

1. An earlier survey of members conducted by the Town Planning Institute in 1962 showed that 60 per cent were employed in local government, 10 per cent in

central government, 5 per cent in New Towns and other public bodies. Of the remainder, 20 per cent worked in private practice and 3 per cent in teaching (Marcus, 1969). Unfortunately I have not been able to find a more up-to-date survey than this.

2. This can be compared to a Gallup poll conducted in 1969 where 51 per cent of respondents supported the Conservatives, 30.5 per cent Labour and 13 per cent the Liberals (Marcus, 1969, pp. 58-9).

3. See, for example, Halmos, 1973; Johnson, 1972; and Volmer and Mills (1966) for general work on professionalisation.

4. 'Social planning' is a highly confusing term. In a sense all planning is social planning in that it has social effects — intended or otherwise. See, for example, Broady, 1968, Ch. 5; Palmer, 1974; Stewart, 1973.

5. An interesting insight into the portrayal of the public in planners' and architects' drawings, where stylised people pose at street corners, gaze at vistas, and elegantly stroll down boulevards, is given by Wood *et al.* (1966).

6. What the observer in the public gallery does not see is undoubtedly much more important, including any agenda items which are discussed privately after the public has been asked to leave, and the discussions of the dominant political group which take place sometime prior to the committee or full council meetings.

7. See, for example, Bailey and Brake, 1975; Corrigan and Leonard, 1978; Cowley *et al.*, 1977; and Jones and Mayo, 1974, 1975.

8. The former Labour GLC required a residential element to be included in office proposals (GLC, *Minutes*, 11 March 1975, pp. 194-7); the London borough of Tower Hamlets also has a public statement (*Policy for the control of office development in Tower Hamlets*), and the London borough of Camden (*A plan for Camden*, March 1977, pp. 12, 13).

9. A directory of some 180 alternative bookshops in Britain, updated to December 1977, is published by J.L. Noyce (1978), P.O. Box 450, Brighton. Counter Information Services and Social Audit may be contacted at 9 Poland Street, London W1, and Labour Research Department at 78 Blackfriars Road, London SE1.

10. For example Goldmann (1969, Ch. 3).

11. As discussed by Bonnier (1972) and Mason (1978), who detail several strategies for the incorporation of local groups by the state.

# 5 THE PLANNING PROCESS AND PUBLIC INFLUENCE

This chapter continues the discussion of the political process of goal-setting and ordering priorities in planning, originally mentioned in Chapter 1. There I emphasised the importance of commercial considerations in establishing what might be built, or to what uses particular land or buildings might be put, and the role of local councillors and professional planning officers in controlling the commercial development process. Here I look at the opportunities open to interested individuals or groups who want to influence land-use planning decisions, and their likely effectiveness.

The issue of public participation in British planning has been an important one in recent years, as described by Damer and Hague (1971) and Thornley (1977). There was growing criticism of planning in the 1960s, partly as a consequence of the development of large-scale housing renewal schemes, many involving high-rise blocks, planned seemingly without reference to the needs and preferences of the people who would live in these new estates. Town planning was condemned increasingly as negative and unduly restrictive, a process subject to long delays and poor results. There was growing disillusionment on the part of the public with the practical outcome of planning policies, and a lack of confidence in the capacity of traditional government institutions to represent and reflect the views of the electorate. People who would be affected by planning proposals were demanding the right to have a say in setting the goals of plans for their area, in deciding priorities, assessing what kind of facilities were needed and would be beneficial, and, conversely, which facilities were not needed and would not benefit local residents. This demand for a more direct involvement in the land-use planning process has parallels in the spheres of industry, education and housing, and can be seen as part of a general demand for a redistribution of influence in favour of powerless groups (Arblaster, 1972; Lapping and Radice, 1968). What is meant by 'participation' may vary from one person to another, and different assumptions have led to considerable confusion in practice. In this chapter I shall look at a range of definitions of participation, and then pay particular attention to the view of government, and the role of local political activity in land-use planning and housing issues as a form of participation in local government decision-making.

## Definitions of Participation

It is useful to begin with some discussion of different definitions of 'participation', in order to separate distinct strands of meaning. This needs to be linked to the earlier general consideration of theories of the distribution of power in society, and I shall refer back to it from time to time.

Various meanings of participation range between tokenism and power-sharing, and there have been attempts to classify them by Arnstein (1969), Dennis (1972) and Thornley (1977), for example. A classification of participation must include the extent to which participation concerns the determination of ends rather then means, or merely influence on means, and shades of 'participation' even below the level of influence — what Castells (1976, p.163) calls 'demagogic' activity. Many discussions of worker participation in industry or participation in planning do not consider the extent to which participation concerns the determination of ends. There are assumptions that what is of interest is the mechanics of involvement, that the purpose of involving people is 'understood', and that all participants have the same understanding of their roles. In itself, this omission has been a source of much confusion.

Different degrees of participation in the workplace could go from complete worker control to participation as a management technique. A range of examples illustrating varying degrees of influence might include a workers' co-operative at one extreme, a work situation where the management is to some extent receptive to suggestions from workers, perhaps on a particular range of issues, or where the management is only receptive to suggestions concerning 'non-controversial' issues, like welfare and social facilities, and the situation where management is not receptive to suggestions from workers at all, at the other extreme. Allowing a few workers to sit on the board, or seeking to be so allowed, is to accept a limited view of worker participation, with premisses largely pre-set. In such situations workers would make similar kinds of decisions to any other board members. Since it is a fundamental tenet of a market economy that firms are in business to make a profit, and must treat their shareholders' interests as paramount, only the means of achieving this objective are open to influence, not the objective itself. To take worker participation to mean token representation on the board is to pay too much attention to *who* makes decisions, rather than *what* decisions are made, or might be made. A similar range of examples from land-use planning would include the implementation of a residents' plan where the underlying values had been discussed and decided by residents at one extreme, the implementation of a local planning authority plan with some 'input' from residents on a predetermined range of issues, and the situation where the relationship between people and planners is confined to giving and seeking information, which is largely ignored except for small, even derisory points, at the other extreme.

A professional view of participation maintains that the only feasible time for the public to be involved in plan-making is after a limited range of options have been 'clarified'. Choices can then be made between a series of set 'packages' and people can vote between them. Questionnaires ask respondents to rank problems or planning issues in order of importance, though, as a rule, what counts as a problem or planning issue has been defined by planning officers and politicians. An enormous degree of control is thus retained by the planning authority, though this point must not be over-stated. An exclusive focus on the 'managers' leads one to overlook the context and constraints within which they operate, as I have argued. An example of this is provided by conflicts of interest over planning goals and strategies in the London borough of Southwark. The council decided to encourage office development on land awaiting redevelopment alongside the River Thames. Private companies have been interested in such development, and it will enhance the borough's rate income, the council argues. Local residents' organisations and trade unionists oppose office development on the grounds that the sites should be put to socially desirable uses, meaning low- and medium-cost housing to rent, industry which will provide skilled manual work, and community facilities. Because of the council's commitment to office development, the only questions that remain are 'How much office space?' and 'What kinds of other development, if any, should be ancillary to it?' The question whether there should be office development in the first place is just not negotiable, and any influence that local organisations might have will therefore be limited accordingly. Conversely, if local groups want to be at all influential, or even keep the lines of communication open, which they clearly do, then they have no alternative but to accept this situation, whether they like it or not. More precisely, the only alternative is not to participate at all.

The scope and mechanics of various degrees of participation will now be considered, drawing on six styles outlined by Dennis (1972, Ch.19). It should be noted that Dennis's formulation suffers from an over-emphasis on the machinery for participation. He never explicitly discusses the extent to which ends are involved, and the implication is that they are not.

### 'Participation' as Receiving 'Benefits'

This is participation in its most derisory form, only in the sense of being on the receiving end, with the role of the local authority merely to inform citizens of its intended courses of action. The public is seen as incapable of expressing opinions and 'participation' takes on a paternalistic, 'welfare-colonial' style. It is a form of manipulation where people are 'served' and 'planned', and it is 'scarcely reasonable that their own definition of their needs should be taken seriously' (Dennis, 1972, p. 256). This style is typical of the operation of local authority social services, housing and land-use planning, though increasingly being challenged.

### Participation as Carrying Out Tasks

Here the 'client' is at his or her most co-operative, doing paid or voluntary work under the control and guidance of the government agency. It is more characteristic of welfare programmes than land-use planning, though the Skeffington report on public participation in planning (HMSO, 1969b) has suggestions about how people can contribute to the plan-making process by carrying out unpaid tasks such as arranging meetings, staffing exhibitions, distributing material and assisting in survey work.[1]Local-authority-sponsored environmental campaigns which utilise the energies of voluntary organisations to clear rubbish and tidy up unused land serve as another example of this.

### Participation as the Dissolution of Organised Opposition.

Here there is agitation for involvement on the part of excluded groups, and 'participation' is used as a stratagem for forestalling or combating opposition by encapsulating the (potentially) disruptive activist or group (Burke, 1968; Selznick, 1949). Etzioni (1968) distinguishes encapsulation of organisations and co-optation of their leaders, a distinction which is employed by Rose and Hanmer (1975) in discussing the squatting movement:

> put concretely the question is not, as it is widely formulated — have Jim Radford and Ron Bailey sold out — but whether squatting as a collective activity became encapsulated? While the two may be connected, it is not a necessary connection and a leadership may retain its radical rhetoric but nonetheless experience encapsulation (p.40).

Similarly Cockburn (1977, Ch. 4) warns that increased opportunities for working people to participate in making decisions which affect them are both positive opportunities, but also carry the possibility of incorporation and control. The obvious advantage of participating is the opportunity this gives for putting across needs and preferences directly, to those officials responsible, in the hope that they will take action accordingly. But there is always the danger that one will buy this participation by making concessions. For example, in an attempt to affect decisions in their favour, so-called 'responsible' squatters may find themselves strenuously dissociating themselves from so-called 'irresponsible' squatters. It is easy to lose one's perspective and become co-opted. One is then unwilling to appear too critical or confrontational for fear of jeopardising one's fundamentally weak participatory status. In this way the carrot of participation becomes a control on people's activities and expectations.

According to Rose and Hanmer (1975, p. 40) Donnison's proposal for local government organisation referred to above has a theoretical counterpart in Etzioni's model of a cybernetic democracy, 'with its apparent participation at the grass roots and control through information at the top'. They suggest that

> when community workers and activists help to create the cybernetic model of social functioning they may be using participation as a substitution for a real redistribution of resources. It may be that participation is the one limitless resource contemporary society is prepared to offer.

Finally, in this connection it is interesting to note the way in which Urban Aid grants to local organisations provide minimal funds to employ activists and keep them occupied, but not sufficient for them to be disruptive.

*Participation as Attention to Consumer Demands*

Here the participant is seen as a consumer to be satisfied and thus is given some opportunity to articulate his or her preferences from time to time. 'Participation as market research' is seen by those responsible for planning and providing services 'as a matter of securing reliable feedback from clients in the form of possibly useful advice and suggestions' (Dennis, 1972, p. 254). The citizen is an informant, not a decision-taker. Techniques of involvement appropriate to this view of participation include surveys and questionnaires, receiving information at public meetings or through exhibitions, making representations to councillors and planners, and so on. How the local planning authority chooses between conflicting points of view, or whether or not they take account of opinions opposed to their own is not mentioned by Dennis, though these are fundamental questions and should be raised.

*Participation in the Decision-making Process*

In this style the citizen is 'no longer a client who is either cossetted or duped'. 'He is cast in the role of policy maker, a voting member of the governing board of directors' (Dennis, 1972, p. 257). Pressure for this kind of involvement usually comes from local groups, and many are trying to secure this type of participation in local affairs, perhaps the best publicised example being Golborne Neighbourhood Council (GNC), in the Royal borough of Kensington and Chelsea (Blair, 1971; Clark, 1970, 1972; *Community Action* (7), 1973, pp.30-1; Gavin and O'Malley, 1977). An important aspect of the Golborne Neighbourhood Council was the fact that it was elected, and one of the very few local organisations in Britain which could not be called unrepresentative, for there was a larger turnout for the GNC elections in 1973 than in the official local government elections. Accordingly, those people elected to the GNC could say with justification that they were more representative of opinion in their area than their ward councillors who sat on the borough council. This put the GNC in a challenging position. A local authority can usually ignore or refuse to meet criticisms voiced by local organisations, if it wants to, on the grounds that they are a minority group whose views do not need to be taken seriously. For the local organisation there is no easy way of proving that they are not merely a vocal minority or of shaking off this label. Bonnier (1972, p. 2) writes about the activities of neighbourhood associations, stressing their 'never-ending search for a recognition of *representativeness* by the residents of an area and by the public authorities'. The city political system has tried to assimilate and co-opt these organisations by the strategies of selective attention, the controlling of subsidies or grants, and allowing them to participate in urban management with conditions, as through an advisory body like a planning forum. Participation in urban management, or in some cases the very existence of the organisation, depends upon its relinquishing any capacity for autonomous action, and furthering the aims of the local authority. Groups which are thus incorporated are rewarded with concessions, often in the form of a monopoly on information, and a role to play in urban management, though not one which challenges the autonomy of the local authority. Mason (1978) describes this process in a Manchester suburb.

Redpath (1973) gives an account of the dialogue between residents of Golborne ward and the Greater London Council (GLC), which recognised the potentially disruptive power of the residents' group — a particularly well organised group, highly politicised and articulate. Levin (1971b, p. 1090) had pointed this out well before the GLC came into the picture:

> It has been said that the Golborne Neighbourhood Council has no powers, and indeed in the formal sense it has not. But as a political force a Council of this kind has considerable potential, not least because the local authority will be unable to ignore it. The authority will inevitably find itself under an obligation to justify its actions in the council's area ... Its strength will lie in its ability to mobilise and unify local opposition.

Redpath describes the organisation as being willing to collaborate with the GLC which in its turn was willing to make concessions, and uses an exchange model to describe the bargaining that went on between residents and officials.

A weakness of the formulation which attributes power to local organisations sharing in local decision-making is the normal interpretation of this ostensibly radical style of participation as participation only in *local* affairs, within a framework of policy and premises already largely set. In general, whether the level is local or wider, the question of the extent to which premises are pre-set is crucial, and must be taken into account. The fact that people's views themselves are constrained by definitions of what is possible, set in terms of the *status quo*, must not be overlooked. They too take account of the 'realities of the situation'. To enter into any dialogue at all one must appear to be 'reasonable', hence much local action either concedes a lot at the outset, or is probably a disillusioning experience.

The limits of local action are well expressed by Coates and Silburn (1972, p. 16). The inherent problem is that although poverty, bad housing, poor schools, inadequate amenities and bureaucratic management systems all 'have a local face' and must be resisted locally,

> they are aspects of one *central* problem whose roots are not at all local but are to be found in the total social system which creates inequality and constantly reinforce it... Community action can only be seen as the first stage of an attack on such problems as these. The second and higher stage is political action, on an integrated scale,

thus raising all the major questions of social deprivation in the heart of the political system.

> No-one should imagine that the struggle to give political expression to the converging anti-poverty lobby will be an easy one, and it is certainly not apparent that the Labour Party can be co-opted into the process. But if it cannot be co-opted, then it must be divided precisely on these questions, because until it is compelled to debate them fully and honestly they will not be able to give rise to any effective overall counter-movement (p. 17).

Purely local action cannot achieve the objectives that some local groups have set themselves, and they are thus directing their activities towards inappropriate targets to some extent. They also need to enter the national political arena, as spelled out by Castells (1976, 1977) and Olives (1976), and as mentioned in Chapter 2, deriving support from similar organisations in other localities, and established groups like trade unions operating in the same locality on different issues, but issues which basically have the same focus. Such co-ordinated political action is not what is usually meant when people talk of participation in planning, though it is a logical development for many organisations and necessary for success. A practical problem with this is that local authority services are often of different standards in different areas and groups are almost bound to have an organisational structure which parallels that of their own authority. This makes co-ordinating activities across local authority boundaries somewhat impracticable and difficult to sustain.

### Participation as Grassroots Radicalism

Dennis's stress on the mechanics of participation has led him to include a sixth style where the 'client' is at his most conflict-oriented. This would apply to many examples of local political activity, where organisations are trying to pressurise the appropriate council, to persuade it to adopt their aims and priorities as council policy. I have discussed the limitations of this formulation above with reference to the activities of the Golborne Neighbourhood Council in North Kensington. Dennis sees this type of activity as a challenge to local bureaucratic domination. However, this focus obscures the fact that much local radicalism is, or in my view should be, directed towards more fundamental economic domination, for the only structure Dennis appears to recognise is an economically empty structure of authority. I return to this point later in this chapter, and consider some empirical examples of local political action over land-use planning issues, and the problems confronting individuals or groups who want to influence council policies which they oppose. Before doing this, however, I want to look at the official government view of participation in planning in Britain, and to locate this view within the range of definitions presented above. For it is the restricted approach to participation adopted by government, with its assumption of a consensus view of politics, which both generates and necessitates this political activity on the part of some local organisations.

### Public Participation: the Official View

In response to criticisms of delay in reaching planning decisions and the general condemnation of land-use planning as negative and bureaucratic, a Conservative government set up the Planning Advisory Group (PAG) in May 1964, to review the land-use planning system, primarily with a view to making it more efficient. The PAG report (HMSO, 1965) recommended a separation of planning decisions into two levels, with only strategic policy plans to be submitted for Ministry approval, and with public participation as an integral part of the process. These proposals were incorporated in the Town and Country Planning Acts of 1968 and 1971, with their provisions for Structure plans and Local plans, publicity and the consideration of comments made to the local planning authority. Meanwhile, the

Skeffington Committee was appointed in 1966, 'to consider and report on the best methods, including publicity, of securing the participation of the public at the formative stage in the making of development plans for their area' (HMSO, 1969b, p. 1), and made recommendations concerning publicity, discussions, public meetings, ways of soliciting views from 'non-joiners' and so on. In practice the situation is ambiguous and partly discretionary. Legislation[2] concerns participation in the plan-making process in the minimal sense of providing for the publication of a proposed plan, and for a public hearing of comments and objections. Moreover, the Minister must be informed of the 'participation exercises' undertaken, and be satisfied that these have been 'adequate', when approving plans. However, planning authorities can decide to a very large extent how much involvement they invite, or tolerate, from local groups and individuals, at what stages, on what issues, and so forth. They can also decide whether or not to take any notice of representations made to them.

Many planners, especially the more senior ones, argue against public participation by referring to their own professional expertise. Land-use planning is a technical activity, engaged in rational problem-solving, with planners taking a broad overall view of the situation. A high degree of training, skill and experience are required, and therefore involvement of the public in anything other than a marginal way is inappropriate, superfluous and time-wasting. After all, 'You wouldn't tell a surgeon how to do an operation, would you?' The demand for public involvement in plan formulation brought the political nature of planning activity into the open. This view of planning diverged, somewhat threateningly, from the one which prevailed, and continues to prevail in professional circles. It resulted in a considerable degree of questioning and self-appraisal on the part of the profession, and discussion as to the purpose and ideology of land-use planning, its area of particular expertise and control, appropriate training for professional membership and so on, which is well documented in the professional journals (Bor, 1970; Grove and Procter, 1966; Keeble, 1966; Kennet, 1968; Lichfield, 1968; (R)TPI editorial, 1971b).

The participation issue has tended to be redefined as a problem of communication, to be solved in the long term by educating people to appreciate the complexity of the planners' task, and in the short-term by exchange of information between local planning authorities and public. Reynolds (1969, p. 144) put it like this:

the quality, quantity and degree of participation is dependent upon education, and the establishment of a simple and effective method of two-way communication incorporating a feedback mechanism...in the short term...the onus is on the local planning authority to present information more readily, and in a way more acceptable to the public.

Evidence submitted to the Skeffington Committee by the professional institutes was very much in this vein ((R)TPI, 1968; RIBA, 1968). Emphasis was also placed on the amount of staff time participation would take up, the problems of blight which would be more acute if public discussion meant that decisions take even longer to reach, the relative remoteness of staff from the public after local government reorganisation in April 1974, and the lack of public scrutiny into highway proposals, which led Levin (1968) to question whether the institutes wanted public participation at all.

As might be expected, reasons for participation in land-use planning given by planners mainly emphasise aspects which will assist them in their managerial role. Planners expect to benefit from residents' detailed knowledge of local areas, acknowledging a lack of detailed information, for many planners have little acquaintance with much of the local planning authority area, and may not live there. Another advantage claimed for participation concerns reducing formal opposition both at public inquiries and in general, thus shortening the time it takes to decide an application. The assumption is that the more people know about land-use planning works, the more they will sympathise with planners on account of their work-load and the difficult decisions involved. This increased understanding is expected to lead to fewer formal objections being made. Then there is also the idea that 'better' decisions will be made, which take account of residents' needs as expressed by the residents themselves, though the implication is that the extent and manner of 'taking account' is for the authorities to decide (Davidoff, 1965).

In contrast to these justifications of participation are two further reasons. Participation is also seen as a means of securing redress against maladministration, as a counterpart to other forms of administrative justice (Dennis, 1972, Ch. 1), and of making planning more democratic, what Rein (1969) has called a 'search for legitimacy', acknowledging that the practice of land-use planning tends to be remote and unaccountable to the public.

The professional view is reinforced by that of government —

indeed the two are in part one and the same — government having accepted a non-political definition of land-use planning, that is non-political in both the 'party political' sense, and the 'conflict of interests' sense. The Skeffington report is clearly in the tradition of the consensus view of politics. There is no consideration given to the possibility that participation may promote conflict:

> there will need to be a new spirit of open discussion of planning issues and a recognition that effective participation will imply a more or less continuous open debate, moving on through the planning process but bearing the seeds of controversy at each stage...but we see the process of giving information opportunities for participation as one which leads to greater understanding and co-operating rather than to a crescendo of dispute (HMSO, 1969b, p.5).

And on the question of consensus on values the report states that the broad aims of planning are implied and accepted in Britain, and do not need debating. It is 'the alternative ways of achieving them' (p. 24) which need to be discussed. As Levin and Donnison (1969) observed, Skeffington made it all seem like a genteel Quaker meeting.

Ten years have elapsed since the publication of the Committee's report, now thought by many planners as a rather naïve over-statement of the necessity for and benefits of public participation, the high point of liberal-mindedness from which it has been necessary to slip back somewhat in order to get on with the day-to-day realities of administrative routine. Public consultation must be put in perspective. It is just another aspect of the job — holding exhibitions, an occasional public meeting, inviting residents to complete questionnaires and so on. 'Participation' has been institutionalised in 'participation exercises' and in this way some of the thrust for public influence has been contained, incorporated and deflected, such that the potential menace to professionalism no longer seems as threatening as it did some years ago. Planners and councillors have successfully maintained their power positions in relation to the public. The land-use planning system has not proved to be as open or penetrable as some people had hoped or others feared. Hoinville and Jowell (1972) suggested that participation was leading towards manipulation by well educated and articulate groups, but the Town and Country Planning Association (1974) went further than this, claiming that whilst 'there is the very real problem of ensuring that the inarticulate non-joiners may have access to the planning system, that system is proving incapable of even utilising the energies of the articulate and enthusiastic.' Skeffington, then, is dead, though there is a legal obligation on local planning authorities to satisfy the Minister that they have engaged in 'adequate consultation', whatever that is taken to mean. In practice, much participation amounts to little more than tokenism.

Participation of a different kind is illustrated by a growing tendency in recent years towards the establishment of new institutions in which universal suffrage is absent and in which members are drawn from various groupings — business, trade unions, local politics and so forth — and participate as social partners. Members of such organisations are not elected from the total electorate, but appointed by government departments or local authorities, or perhaps elected from some limited constituency of interested individuals, who have been officially empowered to do this. Thus they are not accountable to the wider public. These institutions provide a forum for discussion, and in addition may have an advisory or investigative role. Pahl and Winkler (1974), for example, see the development of such institutions as a feature of a corporatist state. Lojkine (1977) argues that it represents a new element on the ideological scene, denying the class affiliation of members and also depoliticising the planning process. He notes this trend in France. In Britain it is exemplified by Neighbourhood Planning Fora set up by local authorities and consisting of elected councillors, planning officers and local residents, the 10 Regional Planning Councils — advisory bodies comprising some 25 members chosen for their wide knowledge and experience of their region but not delegates or representatives of particular interests, and Community Health Councils, with members nominated by local authorities, voluntary organisations and Regional Health Authorities. These organisations provide official channels of communication for their members, and there may be official pressure on local groups to use and work through such bodies, rather than independently of them. They will probably be considered legitimate arenas of debate by officialdom, and as a corollary it may be easier to undermine the activities and 'representativeness' of non-official groups active in the same locality. There is also the possibility that friction between the officially recognised Neighbourhood Forum and other local community activity will sap energy and interfere with the clear articulation of opinion on planning matters.

This raises the question of local political activity, specifically connected to planning and housing issues, to which I shall now turn.

### Participation as Local Political Action

The use of different definitions of participation has led to a confused discussion, and as a result attempts at some sort of dialogue between planners and public have often ended in frustration, neither side feeling that there is a genuine interest in co-operation. Examples of this are legion — the bi-monthly journal *Community Action* quotes literally hundreds of them, to name but one source — and probably virtually every local authority is affected. It is useful to consider a number of well documented cases to illustrate this point.

Studies of local working-class organisations active in attempting to influence local authorities on housing issues — particularly concerning rehabilitation or renewal — have been undertaken by many writers (see, for example, Batley, 1972; Davies, 1972; Dennis, 1972; Lambert, 1975; Lambert *et al.*, 1978; Mason, 1978). The activities of these various organisations have met with mixed success. The local authorities are characterised as large, impersonal, remote structures, seemingly irrational from the point of view of the local residents, disinterested in people's problems, inaccessible and unconcerned. Usually such studies confine their attention to the activities of one particular group and describe the issues involved, the tactics used by the conflicting parties and the outcome.

The selective nature of influence becomes apparent where more than one organisation is involved, and where there is an element of competition between them for councillors' interest and attention, as in Dearlove's study of a range of interest groups in Kensington and Chelsea, the way in which these groups are viewed by councillors, and their ability to influence council decisions. Dearlove shows that councillors make assessments of such organisations as helpful or unhelpful, whether or not their aims accord with council policy, and whether the channel of communication they use is considered proper or improper. There are eight theoretical combinations of these three attributes, but it is suggested that only two are likely to occur in practice.

> The helpful groups had acceptable demands (or none at all) and went about the process of demand presentation (if they were involved in this) in the proper way... The unhelpful groups had demands that were unacceptable, and they invariably had a style

of demand presentation that was regarded as improper (Dearlove, 1973, p. 168).

Influence becomes more difficult to trace with certainty when the
focus of attention is shifted. An example of this is Ferris's account of
the ways in which local organisations have influenced planning
policy in Barnsbury, Islington. A more recent account of the 'Barnsbury scandal' is given in Cowley *et al*. (1977, pp. 178-83). In addition
to the relationship between local groups and Islington Borough
Council, there is the relationship between Islington and both the
GLC and the Ministry, and the short period of Conservative control,
from 1968 to 1971, of a council which had been Labour for at least 25
years previously, and which reverted to Labour control — though
not necessarily the earlier style — in 1971. The two main interest
groups as far as land-use planning is concerned, identified by Ferris
(1972), had very different aims and membership. Both hoped to
bring local authority concern and spending into line with their
particular priorities.

The Barnsbury Association, an amenity society, consisted of a
group of young professional people who had recently bought houses
in Barnsbury. They opposed the council's policy of what they called
'piecemeal redevelopment', and campaigned for the introduction of
a co-ordinated policy of housing rehabilitation, traffic management
and environmental improvement. The Barnsbury Action Group, by
contrast, consisted of local residents, generally working-class people
who had lived in the area for some time, together with shopkeepers
and community workers. They became active at a much later stage,
in an attempt to counter the influence gained by the Barnsbury
Association, and to draw attention to what they regarded as a mis-
allocation of resources on environmental improvements which
ignored the pressing local problem of housing stress. Partly as a
result of the activities of the Barnsbury Association, ten
Conservation Areas had been designated in Islington by April 1969,
with one covering the Barnsbury area. The traffic management
scheme, though amended by the Labour council after 1971 partly as
a result of pressure from Barnsbury Action Group, was introduced.
Ferris attributes this success to members' expertise and familiarity
with acceptable planning techniques like traffic management for the
maintenance of environmental standards in residential areas, as
popularised by Buchanan (1963). Also important was the support
their views were given by Richard Crossman, the then Minister for

Housing and Local Government, and at local level the Conservative election success of 1968, when three Barnsbury Association members were themselves elected to the Council as 'independents'. Ferris's claim that 'had the working class tenants participated on the same level as the Barnsbury Association it seems improbable that traffic planning would have received the emphasis it has' (p. 82) is a moot point, and a comparison of the levels of activity of the two groups partly misses the point. The difference in their aims is extremely important. The goals of the Barnsbury Association for rehabilitation and environmental improvement were viewed sympathetically by the Minister. They were in line with the provisions of the (then recently enacted) 1967 Civic Amenities Act, and the latest planning thinking concerning the impact of traffic in residential areas. They were also viewed sympathetically by a Conservative-dominated council containing three Barnsbury Association members. Against this it seems highly improbable that the Barnsbury Action Group could have made much headway unless they had modified their aims considerably. Their aims were simply not accepted, and they were thus condemned to remain in opposition, their influence limited to amendments being made to the traffic management scheme that was finally introduced.

However, the interrelationship between the activities of the Barnsbury Association and market pressure for gentrification should not be overlooked. The rise in house prices in Barnsbury continued in response to market forces, despite the efforts of the Barnsbury Action Group to influence Islington council policy on housing, and to some extent irrespective of the activities of the Barnsbury Association. Cowley *et al.*(1977) comment on Islington council's 'policies of environmental improvement in an area where property prices were spiralling upwards anyway' (p. 179). Gentrification was obviously in the interests of the middle-class newcomers who had bought into the area. Conversely, their successful campaign for the designation of the Conservation Area, tree-planting and traffic management greatly encouraged the gentrification process.

These studies provide a wealth of interesting material, though I have argued that the bureaucratic perspective is unduly restricted for an understanding of land-use planning conflicts. Dearlove's work is important in its own terms, though he does not emphasise interests which are not active and visible in the political process. These are of two kinds: those to whose interests policy automatically defers without any positive assertion on the part of the interests in question, for example business, and those with contrary interests who never even consider opposing because they feel that it would be pointless. This is crucial, for in these invisible interests and their representation or lack of representation in the assumptions of policy are the hidden parameters of power, as argued, for example, by Bachrach and Baratz (1970) and Westergaard and Resler (1975). Ferris's focus on the two competing local organisations leads him to overlook the role of economic forces in the gentrification of Barnsbury, and hence he tends to over-stress the role played by the Barnsbury Association. He notes the rapid rise in property values but these are not related to wider questions of the encroachment of the middle class on traditionally working-class housing areas, in this case conveniently located near to the City and West End.

A study of activities by people awaiting rehousing by the council from clearance areas in the London borough of Newham under-taken by Dunleavy (1977a, 1977b) has a broader focus, looking beyond the relationship between local groups and the local authority to the wider political and land-use planning system. Dunleavy takes up the issue of council housing, how it is provided and for whom, the relationship between the local housing authority and (systems) building contractors, the significance of the choice of housing form and layout. He looks at a specific protest movement where people were campaigning for alternative rehousing other than the unpopular high-rise blocks in the Labour-controlled borough. Dunleavy argues that in this case an exchange model is inappropriate since the people about to be rehoused had no bargaining power, and were not capable of changing council policy, despite widespread mobilisation and a sympathetic local press. There was no basis for successful protest due to the coercive nature of the relationship between clearance area residents and the local authority housing department.

Analysis of the activities of local organisations from a specifically Marxist perspective is undertaken by Castells (1976; 1977, Ch. 14) and Olives (1976). They are interested in looking at the scope and influence of local organisations and their ability to bring about fundamental changes in the distribution of resources in urban situa-tions. In this connection it should be noted that involvement in local politics in Britain over planning and other issues is a minority activity. On the pluralist view, non-involvement is interpreted as disinterest, or satisfaction with the *status quo*. The system provides opportunities for opinions to be voiced, and it is up to the individual to take advantage of them. This view cannot take into account the very real possibility that many people are apathetic from a feeling of powerlessness, where there seems no point in being active over issues which fundamentally do not appear to be negotiable. In a similar way the issue of mobilisation presents something of a problem for those who adopt a structural approach. I suggested above that Olives's and Castells's emphasis on activities and active people gives an impression that there is a lot of protest activity on the part of the working class, and glosses over problems of mobilisation. The fact that protest does occur is not in dispute, but one of the problems confronting both commentators and political activists alike is why people put up with things as much as they do. However, reasons for non-involvement are not hard to find. There are many claims on a person's time, particularly from family and personal relationships and work. Much work saps energy and interest for other pursuits, and experience at work may reinforce a perception of powerlessness. Potentially interested individuals may not know about the opportunities or procedures formally provided by the system for expressing opinions, or of the existence of local organisations which they might otherwise support, despite official announcements and organisations' recruitment efforts. In addition people may be ignorant of what is at stake: the proposals and counter-proposals. It is often the case in land-use planning matters that the first thing many local residents know about a particular scheme is when all the decisions have been taken and development is about to start, or has already started. Even with time, energy and knowledge some may feel that the effort of political involvement is simply not worth making. The general case against people mobilising is a strong one, and especially so regarding land-use planning.

Planning issues are somewhat abstract, and may seem hypothetical and difficult to understand at a glance, affecting a much wider area than an individual house or street and hence less easy to identify with. This is not to take the position that many professional planners espouse, namely that planning issues are intrinsically complicated and require specialist expertise. No doubt for reasons of professionalism, planners consistently underestimate the capacity of non-specialists to appreciate the principles behind what is basically a large-scale housekeeping exercise, where priorities are decided and resources budgeted accordingly. Much of the apparent obscurity of planning matters results from a lack of information, both of facts and of policy. In order to carry on any dialogue with councillors and planners a local organisation needs information on and knowledge of the management of local government, planning administration and planners' methods of working, thinking and terminology. The dialogue tends to be on the planners' terms, using a style of working familiar to them, and which local organisations have to learn if they are not already acquainted with it. This requires some formality and discipline in discussion which newcomers may find off-putting. To be able to contribute to the discussion they require some background knowledge and confidence in speaking, deterring people who feel themselves insufficiently articulate from taking part. Local organisations need to maintain their own information system. In short, they need to run a parallel organisation to the council's planning department, but on a shoestring, and with limited access to the information available to council and business people. Leaks of information, either openly or otherwise, from sympathetic councillors or officials within the local bureaucracy may be important sources.

Then there is the question of initiatives. In land-use planning decision-making the timetable is set by the developers, local authorities, planning inquiries and so on. Local organisations need to react to applications which have been submitted, decisions of the local authority, or to prepare their case for presentation at a public inquiry. Objections to particular proposals must be submitted by a certain date or in time for a specific committee meeting, otherwise they will not be considered. The amount of time available in which to react varies enormously. It may be several months, as with an appeal to be heard by a public inquiry, or it may be only a matter of days to the deadline for objections when an organisation finds out that a particular planning application has been submitted. This means that leaders of local organisations have to respond to the various proposals according to a haphazard timetable over which they have no control. Short deadlines make any large-scale mobilisation impossible, simply because there is too little time to organise anything.

An additional factor is time, for planning issues are often decided slowly. The negotiating process concerning a large scheme will take months, but more likely years. An appeal may take a further year or maybe more. Interest and concentration have to be sustained over that period, and not many people have the time, enthusiasm and sheer commitment required to be regularly active in local organisations concerned with land-use planning decisions. Typically, as appears to be generally true of political and community organisations, such local groups have a hard core of regular activists, sometimes very few in number, and a larger number of people who are active on a more spasmodic basis or who can be called upon to support particular activities. For those few who are very active there is far too much to keep up with. The information on various planning schemes is often sketchy and not always available. The picture has to be built up like fitting a jigsaw puzzle together. Often negotiations between developers and planning officers take place informally, and some understanding of what is likely to be acceptable is reached long before the proposal surfaces publicly and a formal application is submitted for decision. Thus, some commitment may be entered into, at least at officer level, long before local organisations even know about the proposal. They are constantly trying to keep abreast of events, and at the same time to maintain links with the local community.

Both Olives and Castells consider direct action necessary to achieve change, citing rent strikes, occupations, erecting road blocks and so on as examples. Much action taken by organisations involved in land-use planning issues in Britain has a low profile by contrast, mainly involving meetings with councillors and planners, deputations to the Town Hall, discussions with local people, writing letters, reports, objections to proposals, appearing at public inquiries and so forth. The arguments advanced in the letters and through the meetings may be very telling and well thought out, but letters can be put away quietly in the files, and suggestions made at closed meetings do not necessarily need to be followed up. This raises the question of what direct action people could take. Possibilities would include squatting empty buildings, physically stopping contractors' vehicles from entering a building site, using derelict land for gardens and open space, or building informal structures, occupying the Town Hall or other offices, mass marches, pickets and rallies. Some of these activities would be trespassing, now a criminal offence in Britain under certain circumstances.[3]

Whether by taking direct action people can force a reluctant local authority to change its policies is debatable. Castells and Olives do not doubt its efficacy, nor does Lipsky (1970) in his study of rent strikes in New York, though the rent strikers did not achieve all their demands. The purpose of direct action is twofold: to achieve the immediate objective of securing the land or buildings for a particular use or to prevent demolition, and to secure publicity. It is necessary for an organisation to publicise its aims as widely as possible, as part of its efforts at mobilisation, and as a show of strength. The press, radio and television provide free publicity and reach a very wide audience, but there are two serious disadvantages to this, concerning editorial control and the effectiveness of publicity in achieving the aims of the organisation. In attempting to secure publicity, organisations compete with each other for 'newsworthiness'. Action must be dramatic to appeal to editors, though the issues behind the action may receive shallow or distorted treatment.[4] The slant of the news item or feature programme is decided by editors, so that publicity may be a mixed blessing and may not convey an organisation's case to its satisfaction. Even if the publicity is fair or altogether sympathetic, it may not have the effect of forcing changes from the authorities. Dearlove (1973) reported that the councillors in Kensington and Chelsea thought direct action improper, and accordingly disregarded it and the publicity it attracted. Direct action was a last resort for organisations which had failed to influence the council through what councillors regarded as the proper channels, and was not taken seriously. Dunleavy (1977b) notes that there was extensive sympathetic coverage in the local press concerning the protest by clearance area residents against Newham council's system-built high-rise flats, but this failed to change council policy. He concludes that this was due to the coercive nature of the relationship between the powerless protesters and the local authority in this situation.

Both Olives and Castells are of the opinion that organisations originating at the level of 'collective consumption' need to be linked to wider organisations for success and to avoid the trap of reformism or integration, as suggested above. Local community groups with an interest in influencing the planning process in Britain may have connections with similar organisations in other areas, and with local Trades Councils and ward Labour parties. Such links involve attendance at meetings and discussion of common problems and possible solutions, useful in exchanging and spreading ideas and information, and supportive and inspiring to those actively involved who can learn from each other's experience. It gives a sense of reality to the general nature of the problems that similar organisations are facing separately in their particular areas, and has led to campaigns and publications being sponsored by several organisations.[5] Throughout the country many *ad hoc* groups have formed around various issues, and are faced with the problems of securing resources, mobilisation, dissemination of information, co-ordination and the creation of organisational structures to enable them to do this. A plethora of organisations has sprung up around planning issues, each forming to relate to a particular local authority, and paralleling the local government divisions and structures in a highly fragmented way. Members of community groups may be well aware of the need for links to be established between themselves, trade unions and other campaigns, though contact with other organisations involves yet more meetings and in this sense is an additional burden on activists.

Tenants' associations and community action groups might be expected to press their interests by working through established organisations of the labour movement — the trade unions and the Labour Party. This relationship appears to be weak in practice,[6] even though such local organisations almost certainly include individuals who are union members, Labour Party members and Labour voters.

Several issues appear to be involved here. First, there is the fact that land-use planning is treated as effectively 'non-political' by both planners and the political parties. Hence there is not usually a party division over land-use planning issues at either national or local levels.[7] Then there is the organisation and activity of ward Labour parties. Sklair (1975, pp. 263-4) notes the fragmentary nature of the British labour movement, with sporadic local activity at election time, and over strikes or campaigns around particular issues. Hindess (1971) found that middle-class wards had more active members in ward Labour parties than working class wards, particularly those with higher numbers of unskilled and semi-skilled workers. His argument is that middle-class people have become dominant in the Labour Party, and a corollary of this has been the development of what he calls 'extra-political activity' of 'politically isolated groups' (p. 154). Hence tenants' associations, community organisations and squatters' groups are politically active, but outside the established parties. Crouch (1977) notes this bypassing of established institutions of the labour movement by local community organisations, and Sklair (1975, pp. 268-9) found that tenants' associations involved in political issues were likely to be much more radical than their Labour-controlled councils. A third factor concerns the aims and organisational structure of the trade unions, formed to press for better pay and working conditions for their members, with this traditional purpose predominating in their activities and organisation (Allen, 1960; Harrison, 1960; Richter, 1973; Roberts, 1956). They are not structured to deal with geographical problems and issues, nor do they have sufficient resources to fight on two fronts.

Fragmentation is a feature of union organisation and of the local political scene. It is also a characteristic of the whole government approach to urban questions as evidenced in a number of ways. Planning issues are the responsibility of local planning authorities which are only concerned with what happens inside their own boundaries. Specialisation and division of responsibility between local authority departments, and between local and central government, keep different aspects of the same or related issues in separate compartments. Individual applications for planning permission are treated in isolation, each on its own merits, rather than cumulatively. Legislation relating to land-use planning is highly fragmented, including the Town and Country Planning, Housing, Highway, Public Health, Rent and Local Government Acts. Local organisations develop in order to relate to local authorities responsible for particular services. This high degree of fragmentation makes co-ordination between different local organisations less likely to happen, or to be difficult to organise and sustain.

As far as the effectiveness of public participation is concerned, this depends on local circumstances. Community organisations have successfully campaigned for the introduction of traffic management schemes, rehabilitation of existing housing rather than demolition and rebuilding, the provision of play space and gardens, and the abandonment of motorway and airport proposals. Conservation groups have also been influential, and pressed for the conservation and repair of old buildings, the declaration of Conservation Areas to protect groups of buildings and suchlike. Generally speaking, these gains are all relatively small-scale, though not trivial for the people affected by them. However, it would seem that local organisations generally cannot effectively oppose commercial development when private firms are interested in it, especially if they are supported by local planning authorities.[8] The political persuasion of the council in a particular area may make a difference to an organisation's effectiveness, depending upon what it is campaigning for. A group's resources, aims, determination, strategy and strength all influence its success or otherwise. It is sometimes very difficult for those actively involved and for outside observers to say categorically that their actions have caused particular policies to be adopted. There is always the possibility of coincidence, and some other influential factor which may not be obvious at the time. Examples of such external forces include government policies, for example housing subsidies to local authorities, which at one time were calculated so as to favour renewal rather than rehabilitation, and which have since been set so as to favour rehabilitation rather than new building. Another factor of major importance, as I have argued above, is the buoyancy of the property market, and the interest development companies have in certain forms of development in particular locations. Many proposals for office development were quickly dropped in 1974 in London, due to the collapse of the property market, and coincidental to vociferous protest on the part of local organisations.[9]

Where action does not seem to be successful it can always be argued that things would have been even worse if not for the activity of particular organisations. This reasoning certainly helps to sustain activists. Marxist theorists also subscribe to it, on the argument that the state does not give anything away that is not bitterly fought for. Ultimately it is impossible to know for sure or to demonstrate what might or might not have happened had the campaigns and protests not been undertaken. Dunleavy (1977b) implies that some issues are simply not 'winnable', because the people involved have no bargaining power. In such a situation it is almost immaterial how many protesters there are — there will never be enough, for such groups are not in a position to withhold anything which the authorities want.

Drawing on the arguments presented here and in previous chapters, the theoretical approaches to this issue of public participation can be summarised very briefly. Pluralists see participation as the very stuff of democracy, an indication of the openness of the political system and responsiveness of political leaders to pressure from interested individuals or groups. Reformists, and those who stress the power of public bureaucracy such as Dennis (1972) and Donnison (1973), see participation in local political activity in terms of controlling local officials and councillors, and forcing them to be more accountable to local opinion. For Marxists, much officially sponsored participation is little more than tokenism, containing protest and incorporating campaigning organisations. They are interested in people taking power for themselves in urban situations — the power that they are denied in 'participation exercises'.

This chapter, then, has been concerned to show the scope of involvement of the public in land-use planning decision-making, the different definitions of the term participation, and the official view contrasted with participation as local political activity. In recent years many local planning authorities in Britain have been going through the lengthy process of preparing a Structure plan and Local plans, which involve formal consultation with interested individuals or organisations at different stages of the plan-making process. In the meantime, while these plans are in course of preparation or awaiting approval, planning applications continue to be decided, and much local pressure-group activity is therefore aimed at routine development control decisions as well as broader issues of policy. As I have argued above, formal opportunities for participation are usually restricted to a predetermined range of issues, and take place within a framework of assumptions already pre-set. Further, there is always the problem of how the local planning authority can take account of divergent interests, which are sometimes irreconcilable. For some organisations participation exercises simply do not ask the right questions, as in the example of office development on Southwark's riverside, mentioned above, and for such groups opportunities for formal consultation are clearly only partly satisfactory, at best. They need to campaign for their interests outside or alongside the official procedures. This may not be easy, especially as local planning authorities may argue that the opportunities for participation which they provide are 'adequate'. An organisation's ability to secure such channels of communication depends to a large extent on it being able to provide some useful service for the council, or being sufficiently well organised and mobilising sufficient support for the council to be unable to ignore it.

In this chapter I have concentrated on participation in land-use planning decision-making, which involves a somewhat arbitrary distinction as to what constitutes a land-use planning issue. As I argued above, central government departments, other local authority departments, besides the Planning Department, and private organisations and firms all provide different services and influence planning decisions. Land-use planning is separated administratively from the provision of housing, education, medical facilities, public and private transport and so on. This fragmentation is clearly a source of great frustration to campaigning organisations, though with the reduction in public spending over the past few years as part of the attempt to curb inflation, there has been local pressure-group activity of a more general nature, as with campaigns against the cuts and hospital closures. This limited conceptualisation of what constitutes land-use planning is a theme to which I shall return in the final chapter.

**Notes**

1. For example, a plan for housing renewal prepared by residents of Golborne ward, North Kensington, was published in the *Surveyor*, with the comment, 'perhaps such community involvement in area plans could lead to easing the burden on our already overloaded planning authorities' (Leech, 1971).

2. Town and Country Planning Act, 1968; Housing Act, 1969, in connection with General Improvement Areas; Town and Country Planning Act, 1971, sections 12 and 14; Department of the Environment Circulars, 53/72 and 142/73.

3. Under the provisions of the Criminal Law Act, 1977, section 2. See *Whose Law and Order?* (Campaign against a Criminal Trespass Law, 1978).

4. Distortion may work in favour of an organisation sometimes. Lipsky (1970) suggests that the inflated estimates of the numbers engaged in New York rent strikes reported in the press was a contributory factor to their (partial) success.

5. Examples of such publications are *Lie of the Land*, on the Community Land Act (Land Campaign working party, 1975), and *The Great Sales Robbery* on the sale of council housing (Shelter Community Action Team, n.d.).

6. As it is with Claimants' Unions also (Rose, 1973). Sklair found evidence of much symbolic but little or no concrete trade union support for the rent strikes against the 1972 Housing Finance Act.

7. Examples to the contrary are the attempts of the Labour GLC to limit office development in London subsequent to the publication of its office policy in 1975; also the issue of building and selling local authority housing.

8. An interesting account of the influence of conservationists in rural areas is given by Newby *et al.*(1978). In their study of farming in East Anglia, these authors found that conservation societies had been successful in stopping farmers ploughing up footpaths, taking out hedgerows and burning stubble. This was against the interests of commercially minded farmers with small and medium-sized holdings, but did not clash with the priorities of large estate owners, who often saw themselves as stewards, conserving their land for the benefit and enjoyment of future generations, and already sufficiently wealthy by virtue of their land-holding not to need to farm as rationally and efficiently as possible.

9. This is illustrated in an account of local planning decision-making in Covent Garden, for example. See T. Christensen *Neighbourhood Survival. The struggle for Covent Garden's Future* (Dorchester, Prism Press, 1979).

# 6   CONCLUSION

The purpose of this book has been to consider the role of land-use planning in a capitalist society such as Britain — to look at the scope of the planning system, its apparent aims, the mechanisms by which it operates and its possibilities and limitations. I have argued that though it is ostensibly concerned to control commercial development, land-use planning in Britain can only do this to a limited extent, and in general terms supports the interests of big business and landowners. Accordingly, in the last analysis professional planners also support these interests, if only implicitly, in the course of their work. Moreover, local and national organisations can only make limited headway against them in contesting planning proposals and attempting to influence decisions. This final chapter aims to draw together the main themes running through the preceding sections in two ways: first, in a summary of the usefulness of the theoretical perspectives presented above, and, second, in a brief discussion of the reasons for planning and the problems and opportunities which face urban planning in Britain in the not-too-distant future. Any conception of planning is clearly intimately bound up with a conception of the role and nature of the state, and its relationship to different fractions of capital.

## Theories of Urban Planning: an Overview

As an academic discipline and government activity, land-use planning has drawn on a range of ideas and theories about society in order to help to justify its role and carry out its tasks. These include systems theory, cybernetics, management techniques and strands of political theory. There is no separate specifically planning theory, but the adoption of various ideas and specialisms for what they may have to offer for an explanation of the planning process or a conceptualisation of planning practice.

To help an understanding of the land-use planning system in Britain, I have considered four theoretical approaches to the distribution of power in society, and the strengths and weaknesses of these perspectives can be summarised here. The pluralist approach gives detailed attention to the activities of local and national organisations and interest groups and their aims, tactics and effectiveness in influencing political decisions, which would include land-use planning decisions. The serious limitations of this perspective include the assumption that inactivity can be put down to apathy and disinterest, that government is a weak institution with no line of its own, and that influence is only one-way, from the electorate to politicians. Further, the power of business interests is glossed over, though this is not to suggest that business constitutes an undifferentiated 'bloc', for I noted the distinction between different sections of business, based on their capital. Specifically in the context of land-use planning, national and international firms — both industrial and commercial — are generally successful in obtaining favourable planning permissions. Their interests tend to be accepted and supported by local authorities, but this is precisely what pluralist methodology cannot take into account. By researching organisations which are seen to be politically active, pluralists overlook power relationships which do not usually surface in the political arena, such as is often the case with big business and land-owning interests. Thus, a pluralist perspective involves a spurious distinction between 'the political' and 'the economic' and cannot take account of the underlying relationship between land-use planning and economic factors, which is absolutely crucial to any discussion of British planning. This weakness is shared by those who stress the power of public bureaucracy, focusing on politicians and officials and the authority they hold over the people whom they ostensibly serve. In recent years public bureaucracy has grown in terms of the number of people employed, the diversity of functions undertaken, total public spending, and there has been increased professionalisation of jobs in public administration with a corresponding possibility for mystification and unaccountability on the part of professionals employed by the state. The importance of political and administrative decisions for people's lives and opportunities is acknowledged, but the lack of attention paid to the wider economic context means that this argument may be over-stated. Compared with capital or national government, a local authority has only limited powers. In land-use planning matters this is exemplified by the limited control local councils have over the commercial development process and the location of firms. Hence it is wrong to blame planners for shortcomings beyond their control. Both these theoretical perspectives treat conflicts of interest as operating at the level of politics and administration, without explicit reference to the constraints and powers arising out of the economic context, particularly the highly unequal distribution of capital, both as between different fractions of capital and between owners and non-owners.

The work which I called 'reformist' recognises fundamental inequalities in British society, but the 'solutions' associated with this perspective are reformist measures, aimed at ameliorating the worst aspects of inequality, rather than the abolition of its causes. There is a fundamental contradiction in this approach, between analysis and recommendations, allowing that there will probably be no major changes in the economic and political system, at least in the foreseeable future. Others would argue — including Marxists — that there have been and continue to be fundamental changes in the economic and political system, if only because of the inherent dynamic of capitalism, currently the formation of the monopoly stage of capitalism.

I have argued throughout the preceding chapters that a Marxist perspective has the most to offer for an understanding of the land-use planning system, compared to the other perspectives considered, though it is not without its problems. However, there is a clear integration of political and economic aspects of urbanism. I have mainly concentrated on the structuralist work of Castells and others which has been developed specifically in the context of urban issues, and suggested four problem areas in Castells's work: the question of structural determinism and the extent to which individuals are thought to have scope for influencing events; second, the relationship between local authorities and the central state, and whether or not they are merely outposts of central authority, acting as the state's local agents. Local authorities in Britain do have some discretion in areas such as housing and planning, and the difference in political domination of local councils, for example, can have an influence on local authority activity and attitudes. Third, Castells over-emphasises struggle, and gives the impression that there is considerable conflictual political action over urban planning issues, which is not the general case in Britain. The fourth problem concerns a theory of social change, whereby Castells attributes significance to political activity in the sphere of consumption, rather than in the sphere of production, and implies that it is this struggle over urban issues such as housing and planning which constitutes the major impetus for social change. Clearly this emphasis overlooks the much more systematic work-based activity organised by trade unions. Of course, the two spheres of production and consumption are intimately related, though this is not given much explicit recognition in the activities and preoccupations of the left-wing political parties in Britain, which tend to concentrate almost entirely on 'the struggle at work'.[1]

Implicit in my earlier comments on the four theoretical perspectives are fragmented remarks and questions concerning the role and powers of the state and the relationship between the state and capital. This needs to be explored fully, both in general terms and in detail specifically in connection with contemporary Britain, which is clearly beyond my scope here. This focus has been attracting considerable interest in recent years from Marxists and non-Marxists — commentators, officials and political activists — with the general recognition of the worsening international economic crisis. Important questions — both theoretical and practical — concern the extent to which the state can continue to produce the conditions under which capitalism can go on expanding, and the precise relationship between the various component institutions of the state apparatus and different fractions of capital in specific situations. Holloway and Picciotto (1978, Ch. 1) argue that these questions have been dominated by the Poulantzas-Miliband debate in Britain, an argument based on 'a false polarity' in their view, both writers focusing, at least implicitly, on the 'political as a separable area for study' (p. 3), which has led Clarke (1977) to call Poulantzas a 'bourgeois sociologist'.

Wright (1978, Ch. 3), for example, stresses that the capital accumulation process generates contradictions, rather than once-and-for-all solutions, which means that at different stages of capitalist development this process faces different constraints. Solutions to these impediments at any given stage generate the new impediments which constrain the accumulation process at a subsequent stage. He suggests that several structural solutions to the accumulation impediments of advanced monopoly capitalism are now emerging. At the international level this involves foreign investments being increasingly based on manufacturing, rather than the extraction of raw materials, agriculture and trade. Writing in the context of the USA, he notes the recent reductions in public spending affecting welfare programmes and public services. Wright argues that qualitatively new forms of state intervention are called for.

It is no longer enough for the state simply to set the parameters for capitalist production by regulating aggregate demand, interest rates and taxes, and to deal with the social costs generated by the irrationalities of capitalism through police, pollution control and mental hospitals (p. 177).

The state needs to become directly involved in the rationalisation of production and to achieve this will have to increase its ability to control both capitalists and workers. It particularly needs to be able to prevent the flight of capital in the face of increasing state involvement, and to eliminate unproductive sectors in the interests of increasing the productivity of capital as a whole. It also needs to be able to constrain wage demands. He acknowledges that the political obstacles to such increased powers are high. Indeed, he suggests that it will become increasingly difficult in the USA to apply allegedly 'neutral' market criteria to production, and that political criteria will become more central.

Recent German work (introduced by Holloway and Picciotto, 1978) is of interest here, where the form, nature and activities of the state are seen as part and parcel of the capital accumulation process, and the aim is 'to "derive" the state as a political form from the nature of capitalist relations of production, as a first step towards constructing a materialist theory of the bourgeois state and its development' (Holloway and Picciotto, 1978, p. 2). Through its various component apparatuses the state responds to the needs of capital, and hence state activity can be seen as a source of counter-tendencies to the tendency for the rate of profit to fall. Various writers have begun to address themselves to this task, stimulated to understand the role of the welfare state, and its increasing repressiveness in West Germany. Hirsch sums up several strands as follows:

The compulsion imposed on the state to provide, on an increasing scale as the socialization of production increases, decisive material and organizational pre-conditions for the process of social production and reproduction (which is determined by the movement of capital) is certainly an essential basis for 'welfare state illusions' of reform. But this tendency is thoroughly ambivalent politically. When the decline in the rate of profit and the tempo of accumulation becomes manifest, this must lead to an intensified exploitation of labour power mediated through the state apparatus, while at the same time potential state resources for 'superfluous' measures of pacification and reform — 'superfluous', that is, for the immediate profit interests of capitals — are drastically restricted. This is the context in which the 'consequences' of economic growth — decay of the cities, chaotic traffic situation, collapse of the ecological equilibrium, etc. — become politically explosive: not because the 'managing capacity' of the state is too small in a technical sense or indeed restricted by an outdated 'view of the world', but because capital comes up against the self-produced barriers of its valorization, which can be broken through only by an intensification of exploitation and class struggle (in Holloway and Picciotto, 1978, pp. 104-5).

It is in this light that one can understand the development of public intervention, which brings me to the various reasons put forward to justify it, and specifically why land-use planning is said to be necessary.

**Why Plan?**

There are several reasons put forward to justify public intervention in the form of land-use planning. It is claimed that planning the use of scarce resources is more rational than relying on market forces to meet demand. This is both a humanitarian and a pragmatic argument based on the belief that everybody's fundamental needs should be provided for. It is advanced by both liberals and Marxists, and those George and Wilding (1976, Ch. 3) have called 'reluctant collectivists', such as Keynes, Beveridge and Galbraith, who claim validity for market principles, but who nevertheless support some degree of public intervention in the economy to compensate for a lack of rationality in the market, and to provide and co-ordinate services such as housing, education and health care for low-income groups, or unemployment payments, which the market left to itself would not supply. Second, public intervention, including land-use planning, has the potential for some redistribution of scarce resources, with the possibility of reformist compensatory policies leading to a measure of social equality. Then there is an ecological argument, concerning the conservation, wise use and management of scarce natural resources, not spoiling or polluting the environment, and safeguarding resources for future use and enjoyment. Fourth, the introduction of public intervention into a market-based economy may be seen as a way of supporting and facilitating a capitalist system to continue, as I have argued previously. All four justifications for public intervention, and specifically land-use planning, are currently held to some extent in Britain, though the limitations of the planning system mean that possible recommendations arising out of these views are correspondingly restricted in practice. These various reasons highlight the contradictions inherent in the capital accumulation process, and in state intervention into that process.

Perhaps the most obvious consequence and manifestation of the accumulation and centralisation of capital is urbanisation, for large urban areas both house and require for their maintenance a considerable labour of productive and service workers. Land is a commodity in limited supply, and the use of particular parcels of land needs to be decided. The private ownership of land and capital in Britain, and the prevailing profit motive, means that competing land uses tend to be adjudicated in favour of the most profitable use. Thus, in urban areas, industrial and commercial uses tend to take precedence over residential development and welfare and leisure facilities; commercially viable residential development, leisure and welfare facilities take precedence over those that are not commercially viable. Newby *et al.*(1978, Ch. 6) show how councils in Suffolk, a predominantly rural county, act in support of farmers and conservationists by severely restricting industrial and commercial development. This has the effect of conserving the countryside and limiting alternative means of livelihood for agricultural workers.

As the capitalist mode of production has developed, there has been an expansion of urban planning on the part of the state, both constraining absolute freedom of land use for individual households and private firms, but also supporting new urban development by making the initial investment in infrastructure, including transport, communications, water supply and drainage, which tend not to be profitable to supply, and hence of little interest to commercial development companies. There is also continued state provision of 'unprofitable' services, such as housing, education and training, necessary for maintaining a skilled labour force, and grants and incentives to firms to relocate or employ labour. Thus land use is not based on the needs of working people, or any relatively low-paid people, a fact which manifests itself in the increasingly high cost of inner-city housing, and transport costs to the central area from low- and medium-housing areas.

London, like many British cities, has been losing population as inner-area sites are redeveloped or rehabilitated at lower densities than previously, as suburban building has increased, and as non-residential uses, particularly offices, have encroached upon and displaced the residential use of existing buildings. This loss of population from inner areas is sometimes said to be a result of popular choice (Thorburn, 1970), though Williams *et al.* (1972, p. 26), in their study for Westminster City Council, put the counter-view, claiming that 'loss of population is due less to natural choice in favour of the suburbs, and more to external pressures on the housing market.' Land costs have increased beyond the level at which it is economically possible to provide low-cost accommodation in inner-city areas. This also applies to local authority housing, which has generally not been built on high-priced central sites. Writing about housing needs in London, Glass and Westergaard (1965, p. 5) point out that, in addition, 'most of the new private housing has been located at the outer periphery of Greater London, and has been provided for owner occupiers, not for tenants.' Peripheral development on virgin sites is usually less costly than redevelopment within the city (Stone, 1972), and other things being equal, this tends to mean that the bulk of new development proceeds at the periphery.

This is exactly the contradiction outlined by Castells and referred to earlier, concerning the need for low- and medium-cost housing in the city, but the lack of motivation of capitalists to be involved in this. There are basic services on which commercial and business concerns rely heavily, involving such people as maintenance workers, security staff, cleaners, catering staff, public transport operators, local authority cleansing department employees, post office and fire department workers and so forth. Very many of these jobs involve night work or shift work. Employers interviewed by Williams *et al.* mentioned low pay, difficulties of travel in terms of cost, time and effort, and the high cost of living in the central area as the major contributory reasons for their acknowledged difficulty in getting staff. A government report on the manning of the public services in London (HMSO, 1975b) anticipates continued labour shortages in London's public services, and increasingly poor provision as a result. This fundamental contradiction between the need for workers and the difficulty of providing housing that they can afford is also clearly acknowledged in the Greater London Development Plan (Counter Information Service, 1973, pp. 52-8). Eversley, a former Chief Strategic Planner for the Greater London Council, pinpoints what he sees to be the major problem as 'that of the contrast between rising urban costs and falling urban incomes' (1972, p.351). Rising costs include land prices, construction costs, higher fixed costs *per capita* when population is falling, and debt charges. These are to be met out of a falling income caused by a loss of population, especially in inner boroughs, and the higher proportion overall of upper- and middle-income people who are moving out. Such phenomena Eversley claims to be 'part of the normal historical process of city development' (p. 357) which have been 'solved' in the past by the city retaining its population, and therefore its source of income, by a series of boundary adjustments involving the incorporation of the new peripheral low-density housing areas. It should be noted that this 'normal historical process' is 'normal' in the sense that it tends to accompany capital accumulation.

Lamarche, Lojkine and Castells, cited above, all made a distinction between different fractions of capital, and Lefebvre also utilises a similar distinction in his explanation of the booms and slumps in property development:

> Urban redevelopment plays the part of a secondary process parallel to that of industrial production. It is a compensating process: when the surplus value created by industry sinks to a low level, surplus value created by construction and speculative development rises instead (cited in Ambrose and Colenutt, 1975, p. 9).

> capital investment thus finds a place of refuge in the real estate sector, a supplementary and complementary territory for exploitation (Lefebvre, 1976, p. 34).

Certainly, the boom in office development from 1967 to 1974 in Britain, particularly in London, appeared to coincide with stagnation and recession in the economy generally, which has subsequently also affected property development. As a sideline on this, the growth of planning bargaining in the early 1970s, referred to in Chapter 1, is interesting. This occurred during a property boom, where planning authorities were able to negotiate some additional benefit such as provision of a particular facility, use of land, or financial contribution not contained in the original planning application and not normally financially advantageous to development companies. An argument in favour of planning gains offered in the press involved the belief that residents should share in the profits made from redevelopment. Another interpretation of the deals between planning authorities and development companies is to see them as a reflection of a growing economic crisis at the same time as a property boom. At a time when state expenditure was over-extended, and subsequently reduced, many local authorities made gains through planning bargaining, and negotiated the use of private land for public open space, financial contributions for the provision of housing, buildings for community use and suchlike — facilities which they would otherwise have had to finance themselves.

## Problems and Opportunities

The logic of the capital accumulation process means that demands are supplied, rather than needs being met, as has been pointed out by many writers (for example Allaby, 1978; Blackburn, 1972; Gorz, 1977; Kapp, 1978). Decisions made by private firms need not take account of social requirements or costs, merely market demand. Collective needs such as housing, public transport, nurseries, schools, health facilities, sports centres and suchlike are of limited interest, for they stand in a contradictory relationship to capitalist logic. Kapp argues that decision-making in business enterprise has an in-built tendency to disregard those negative effects on the environment 'external' to the firm. It ignores costs which can be shifted to others or to society in general, such as noise, pollution, congestion and commuting, for example. Though public authorities in Britain have some planning powers, a solution to the problem of controlling private development and providing services that people need and can afford would require radical change in the present system, such that government becomes a 'real planning agency'. Ambrose and Cole-nutt (1975, pp. 164-77) give a very full list of changes they consider necessary to achieve this, suggesting that there should be a national building plan; development should be administered by public development corporations; government should take control of financial institutions; all land and property should be taken into public ownership; the basis for commercial rents should be restructured; the power of the property professions — particularly surveyors and valuers — should be reduced; planning should be based on compensatory principles, producing a progressive rather than regressive redistribution of resources; there should be political appointments to the Civil Service and local government, and genuine public involvement in planning decisions; the study of urban issues should be included in school and college curricula, and research effort and results should be used more effectively. These recommendations arise out of the authors' study of the British land-use planning system, and involve both relatively detailed matters concerning present planning practice as well as fundamental changes in the political and economic system. Broadbent (1977, Ch. 7) similarly recommends national planning and nationalisation of land, with the state taking a leading role in the economy, so that state planning is able to achieve national social and economic goals directly. He claims that Britain has a mature economy, with no room left for extensive new growth or exploitation of new resources. The private sector has increasing difficulty maintaining investment profit, and is apparently incapable of halting the overall decline in the national economy. He suggests that there should be a leap from a market-based, market-led economy to a public-sector-led economy, and claims that the 'uncomfortable choice is slowly but inexorably coming closer' (p. 246). Strengthening the planning system would mean calculating costs on an entirely different basis, not merely private sector financial costs. This view echoes that of Kapp, who argues for a radically new form of decision-making and planning, where an evaluation of costs and benefits of investment decisions would be undertaken from a broader and more comprehensive macro-economic perspective, allowing for an assessment of the implications of alternative technologies for the maintenance of human life.

Detailed suggestions may vary from one author or activist to another, and it is not my purpose to present an exhaustive list. It is the principle which is important. As long as there is a private market in land and property, and private initiative in development as at present, control of the commercial development process by both central and local government will suffer from severe limitations. State intervention will be necessary to fund unprofitable development, to provide low- and medium-cost housing, job opportunities and training to ensure some correspondence between the skills that people have and the jobs available. But this reinforces the fiscal crisis of the state, as stressed by Yaffe (1973) and O'Connor (1973), for example, who emphasise the role of state expenditure in the present crisis. Hence public expenditure is not a cure for the ills of capitalism, but the source of yet another dilemma and contradiction.

However, the very many suggestions for change which have been made repeatedly have not been implemented, for this would require fundamental political and economic change.[2] A real planning agency would also involve a much more broadly based view of planning resources, compared to the limited land-use planning currently practised in Britain. This can be illustrated by two examples. First, an issue which currently faces British society and which is unlikely to be solved by conventional thinking and methods is that of unemployment. I noted above the rise in unemployment levels, their marked regional effects and the various subsidies that the state is paying to create jobs or to avoid redundancies. Some of this unemployment is caused by structural changes in the economy, such that jobs are being created in the tertiary service sector but being lost in the primary and manufacturing sectors, due to the nature of the capital accumulation process. The traditional land-use planning solution for areas of relatively high unemployment is to try to attract industry to such locations. However, with the development of micro-processors, the robot factory is no longer a thing of science fiction, but a feasibility (Forester, 1978a, 1978b; HMSO, 1978; Taylor, 1978). Unless the government decides to stifle such developments and forbid their adoption, which in view of what I have been arguing about its involvement in the capital accumulation process is highly unlikely, they can be expected to have an important effect on the employment structure, for fewer employees will be needed. Work is a key institution in British society, in terms of economic necessity and social standing. Those who do not work are often divided into two groups: the 'legitimately' unemployed such as the sick, handicapped and single parents, and the so-called 'work-shy', said to be living a life of ease and irresponsibility. Welfare state provision for the unemployed is calculated so as to enable people to survive an emergency, assumed to be short-term. Unemployment is seen as a temporary, stigmatised phase, and hence payments are set at a notional subsistence level.

The contradiction here is glaringly obvious. As Marx (1954, Ch. XXV, sections 3 and 4) noted in his discussion of surplus population and the reserve army of labour, an inherent feature of a capitalist mode of production is that it does not have the capacity to absorb all those who are potential wage-earners — currently most people between the ages of 16 and 60 or 65. Yet receiving unemployment benefit is thought of by many individuals and the Establishment as irresponsible and shameful. Increased redundancy as a result of automation and a growing awareness of semi-permanent unemployment in some areas will make this contradiction less credible. Indeed, everybody who is potentially part of the labour force does not need to work, with the crucial proviso that there must be some politically acceptable way of sharing out wealth which does not penalise those who do not work. This is a stumbling block that has caused trade unions to try to resist automation in Britain, because of the loss of jobs involved.[3] Similarly, they oppose work-sharing arrangements, early retirement and the introduction of shorter working hours, because their members would suffer financially as a result, through lower pensions or wages.

However, it is not necessary to stretch out the work so that there is enough to go round, but it is necessary to share out the wealth. As long ago as 1932, Russell (1935) was advocating drastically reduced working hours, arguing that the modern technology of that period had made it possible to 'diminish enormously the amount of labour required to secure the necessities of life for everyone' (p. 15). More recently, Illich (1978), for example, has argued for the right to 'useful unemployment'. He is scornful of the need for more and more commodities, consumer goods and comforts, and argues that people in advanced industrial societies suffer from 'modernized poverty',

> the experience of frustrating affluence that occurs in persons mutilated by their reliance on the riches of industrial productivity. It deprives those affected by it of their freedom and power to act autonomously, to live creatively; it confines them to survival through being plugged into market relations (p. 8).

The concept of 'need' is a notoriously difficult one to define, because beyond the basic needs people have for bare survival, other needs are culturally conditioned, with an important element of contrived demand promoted by advertising and obsolescence as part of the expansionist 'logic' of the capitalist mode of production. Both Russell and Illich are arguing for a relatively static level of material needs and consumer goods, which brings them right up against this expansionist thrust. Their approach is based on an optimistic belief in human potential for personal development and enrichment, where people are no longer slaves to machines and boring, routine work, but are able to spend much more of their time more positively. On this view, land use and land and property development would be based on needs, rather than profitability, though there are clearly problems involved in defining 'need' and deciding priorities. The use of resources would include people's right to livelihood, shelter, education, health and amenity on an equitable basis, the wise use of raw materials based on ecologically sound interrelationships and a minimum of waste, and safe food production involving rational price structures.

Another example of a wider perspective for planning can be given by looking at illness and health, for many problems which manifest themselves as medical problems in this society may have other causes. Bad housing conditions, especially dampness, affect health, worsening bronchial, chest and lung conditions, for instance, particularly in old people and small children. There is also evidence of the effect of overcrowding on infectious diseases and mental illness.[4] Housing availability and allocation policies which tend to separate married children from their ageing parents are also important. Hospitals accommodate many elderly women suffering from depression, often brought on by widowhood. After years of marriage, the loneliness they face is unbearable, and the prospect of living alone may be so daunting that they become ill. On the other hand, some elderly people are kept in hospitals simply because there is no suitable housing, such as sheltered housing or a ground-floor flat, available for them. Moreover, people in psychiatric hospitals may not be discharged because there is a general shortage of one-person accommodation, or because they would benefit from a more communal form of housing which hardly exists. Mental illness is on the increase, and over 40 per cent of the hospital beds in Britain are allocated for psychiatric patients at present. Many women suffer from depression; 16 times as many women are treated for it as men, and more working-class women than middle-class women. This very often stems from poor housing conditions, isolation, feeling imprisoned at home with small children, finding it difficult to cope with rising prices and suchlike (Brown and Harris, 1978, especially Ch. 12). Explanations of depression in women offered by feminists stress the fact that women are treated as second-class citizens in a patriarchal society. Family doctors cannot deal with these causes, but can only alleviate the symptoms by prescribing anti-depressant drugs. Several studies have highlighted problems of living in high-rise housing, particularly stressing the lack of opportunities for children's play, the damaging effects this may have on their development, and the associated strain this has on mothers (Darke and Darke, 1970; Jephcott and Robinson, 1971). Modern illnesses such as heart attacks and cancer are aggravated by the pace of life, worry, pressure at work, too little exercise and so on. Additional features of the physical environment which have an adverse influence on health are atmospheric pollution, unhealthy working conditions and road traffic, as noted by McKeown (1965, pp. 81-4), for example.

Such factors fall well outside the scope of land-use planning as it currently operates in Britain, yet are indirectly planning matters. Housing design, provision and allocation should be within the ambit of a more widely defined urban planning. However, such a comprehensive view of planning would require administrative and budgetary co-ordination on a scale never before adopted in Britain, even in wartime. At present, housing, highways, health, social services, employment, education and so on are conceived of in watertight compartments, organised separately, either by various government departments and agencies at central and local levels, or by private organisations and firms. Within government service there is a high degree of professional specialisation. Just as the majority of land-use planners do not consider health matters to be part of their concern, conversely members of the medical profession do not feel that it is part of their role to take a stand over housing conditions or income levels. However, the fundamental difference in an integrated planning system should be in approach, organising resources in the interests of people, and with their needs predominating, rather than for profit and capital accumulation. This overall objective of people's health, welfare and scope for creative living would be the guiding principle for devising appropriate organisational structures. Resources could be made available to groups of people or on the basis of geographical areas, to be used according to their priorities and needs.

Allaby (1978) and Robertson (1978), for example, sketch out possible scenarios of this kind for Britain, incorporating elements of intermediate technology, self-sufficiency, local food production, recycling, some return to craft methods, reduced energy consumption and so on. Pahl (1978) makes similar suggestions for people on low and fixed incomes, such as pensions or unemployment payments, in inner-city areas. There are two underlying strands to such suggestions. These writers assume that the problems generated by the world economic crisis and rising oil prices will have serious repercussions on life in Britain, and reduce the material standard of living for years to come. However, despite these apparent constraints, they argue that the quality of life could be greatly improved for very many people if resources were allocated and used in a different way.

Increased unemployment and cuts in public spending aimed at curbing inflation and affecting health, education and social services particularly are already with us, and can be expected to get worse. The present Conservative government's first Budget, with income tax concessions favouring those in the top tax bracket and a rounding-up of VAT which will raise prices for everyone, is another instance of increasing exploitation. As long as the driving force behind commodity production is private profit, there can be no resort to environmentalist solutions unless these will advantage capitalist interests. For example, the increasing price of oil is forcing industrialised countries to consider alternative forms of energy. Some firms and government agencies are beginning to explore the possibilities of intermediate technology or recycling re-usable materials. Increased unemployment is not only demoralising for some individuals, but also a possible threat to political stability, which, with falling living standards, cannot be ruled out entirely, even in Britain. Further cuts in public spending mean that, under pressure, government may be prepared to allow access to currently unused housing, and increasingly support housing co-operatives and tenant management or sale of housing estates. It may allow people to make use of otherwise derelict land for gardens and playgrounds, or unused canals for leisure pursuits. The work of establishing and maintaining these facilities is thus transferred to voluntary associations and interested individuals; substantial economies in public spending are made, and amenities provided largely as a result of people's energy and determination. To a limited extent these things are already happening in various parts of the country, though so far no more than straws in the wind.

One of the dilemmas for the state is that while facilitating the capital accumulation process, public intervention in a market-based economy, including land-use planning, is limited by the need not to encroach upon the profitability of the private sector. It has been the burden of my argument that the contradictions and tensions inherent in this process between economic expansion and providing for people's welfare continuously generate new problems and opportunities. Despite the need for housing which people can afford, and social facilities of all kinds, and despite the range of technology which could be utilised to this end, Robertson's 'sane alternative' will not be adopted in any general way until it is beneficial to the capital accumulation process.

## Notes

1. What recognition there is of this relationship between home and work seems to have been introduced by writers from within the women's movement, or others who have been stimulated and challenged by this work, and have to some extent been forced to reconsider their position. See, for example, James and Costa, 1972; Mitchell, 1971; Zaretsky, 1976.

2. By contrast, recent proposed reforms of the planning system are very slight. The Dobry report (HMSO, 1975a) on development control procedures aimed at shortening the time taken to decide planning applications (Ambrose and Colenutt, 1975, pp. 66-7), though these recommendations have not been adopted so far. The Community Land Act, 1975, gave local authorities the power to acquire land compulsorily for 'relevant development' (see Land Campaign Working Party, 1975). The Development Land Tax Act, 1976, aims to deal with restoring increases in land values to the community by taxing profits from increases in land values due to development or change of use. However, Massey and Catalano (1978, Ch. 8) have argued that this legislation will make no serious inroads into the power of land-owning capital. In any case the present Conservative government intends to make three changes which will greatly facilitate the development process and benefit development companies and landowners, namely the abolition of Office Develop-ment Permits, the repeal of the Community Land Act, and the introduction of a lower rate of development land tax.

3. Elsewhere, for example Italy, Japan and USA, a greater degree of automation has already been adopted, for instance in car manufacturing.

4. As a result of their professional experience, many doctors do not hesitate to attribute some illness, at least in part, to poor housing conditions, though systematic data are hard to find. A serious problem for research is that of separating out inadequate housing from other closely related factors such as low income, high unemployment rates and poor diet, which also affect health.

# REFERENCES

Aaronovitch, S. 1961. *The Ruling Class*. London: Lawrence and Wishart

Albrow, M. 1970. The role of the sociologist as a professional: the case of planning. In P. Halmos (ed.), *Sociological Review Monograph*, No. 16, pp. 1-20. Keele: University of Keele

Allaby, M. 1978. *Inventing Tomorrow*. London: Sphere Books

Allaun, F. 1972. *No Place Like Home: Britain's Housing Tragedy (from the Victim's Point of View) and How to Overcome it.* London: Andre Deutsch

Allen, V. 1960. *Trade Unions and the Government*. London: Longmans

———, 1977. The differentiation of the working class. In Hunt (1977), pp. 61-79

Alonso, W. 1964. *Location and Land Use: Towards a General Theory of Land Rent*. Cambridge, Mass.: Harvard University Press

Althusser, L. 1969. *For Marx*. London: Allen Lane

Althusser, L. and Balibar, E. 1970. *Reading Capital*. London: New Left Books

Altshuler, A.A. 1965. *The City Planning Process: a Political Analysis*. New York: Ithaca Press

Ambrose, P. and Colenutt, B. 1975. *The Property Machine*. Harmondsworth: Penguin Books

Amos, F.J.C. 1973. Comment on Alfred Wood's Paper 'City Planning in the United Kingdom'. *Journal of the Town Planning Institute,* vol. 59, no. 6, p. 262

Arblaster, A. 1972. Participation: context and conflict. In G. Parry (ed.), *Participation in politics*, pp. 41-58. Manchester: Manchester University Press

Arnstein, S.R. 1969. A ladder of citizen participation. *Journal of the American Institute of Planners*, vol. XXX, no. 4, pp. 216-24

Ashworth, W. 1954. *The Genesis of Modern British Town Planning*. London: Routledge and Kegan Paul

Bachrach, P. and Baratz, M.S. 1962. Two faces of power. *American Political Science Review*, vol. 56, no. 4, p. 948
___, 1970. *Power and Poverty*. London: Oxford University Press
Bailey, R. 1973. *The Squatters*. Harmondsworth: Penguin Books
___, 1977. *The Homeless and the Empty Houses*. Harmondsworth: Penguin Books
Bailey, R.V. and Brake, M. (eds.) 1975. *Radical Social Work*. London: Arnold
Banfield, E.C. 1961. *Political Influence*. New York: Free Press
Baran, P.A. and Sweezy, P.M. 1966. *Monopoly Capital*. New York: Monthly Review Press
Batchelor, P. 1969. The origin of the Garden City concept of urban form. *Journal of the Society of Architectural Historians*, vol. 28, no. 3, pp. 184-200
Batley, R. 1972. Non-participation in planning. *Policy and Politics*, vol. 1, no. 2, pp. 95-114
___, 1975. *The Neighbourhood Scheme: Cases of Central Government Intervention in Local Deprivation*. London: Centre for Environmental Studies, Research Paper 19
Beckerman, W. 1974. *In Defence of Economic Growth*. London: Cape
Bell, C. and Bell, R. 1969. *City Fathers. The Early History of Town Planning in Britain*. London: Cresset Press
Bell, C. and Newby, H. 1971. *Community Studies: an Introduction to the Sociology of the Local Community*. London: Allen and Unwin
Bell, D. 1960. *The End of Ideology*. New York: Free Press
Benington, J. 1976. *Local Government Becomes Big Business*, 2nd edn. London: Community Development Project Information and Intelligence Unit
Berry, F. 1974. *Housing: the Great British Failure*. London: Charles Knight
Beveridge, W.H. 1943. *The Pillars of Security*. New York: Macmillan
___, 1944. *Full Employment in a Free Society*. London: Allen and Unwin
Blackburn, R. 1967. The unequal society. In R.Blackburn and A.Cockburn (eds.), *The Incompatibles: Trade Union Militancy and the Consensus*, pp. 15-55. Harmondsworth: Penguin Books
___, 1972. The new capitalism. In R. Blackburn (ed.), *Ideology in Social Science*, pp. 164-86. London: Fontana/Collins

Blair, T. 1971. Golborne: Beginning of a Revolution? *Official Architecture and Planning*, vol. 34, no. 5, p. 371

————, 1973. *The Poverty of Planning*. London: MacDonald

Bonnier, F. 1972. The practices of neighbourhood associations and the process of co-option. *Espaces et Société,* nos. 6/7. Translated C.Lambert, 1974. Mimeo

Bor, W. 1970. Presidential address. *Journal of the Town Planning Institute*, vol. 56, no. 9, p. 380

Borja, J. 1977. Urban movements in Spain in Harloe (1977), pp. 187-211

Bottomore, T.B. 1964. *Elites and Society*. London: Watts

Broadbent, T.A. 1975. *An Attempt to Apply Marx's Theory of Ground Rent to the Modern Urban Economy*. London: Centre for Environmental Studies, Research Paper 17

————, 1977. *Planning and Profit in the Urban Economy*. London: Methuen

Broady, M. 1968. *Planning for People*. London: National Council of Social Service

Brown, G.W. and Harris, T. 1978. *Social Origins of Depression. A Study of Psychiatric Disorder in Women*. London: Tavistock

Buckley. W. (ed.) 1968. *Modern Systems Research for the Behavioural Scientist*. Chicago: Aldine Press

Burke, E.M. 1968. Citizen participation strategies. *Journal of the American Institute of Planners*, vol. 34, no. 4, pp. 287-94

Burnham, J. 1943. *The Managerial Revolution*. London: Putnam and Co.

Butterworth, E. and Weir, D. (eds.) 1970. *The Sociology of Modern Britain*. London: Fontana/Collins

Cadman, D. and Austin-Crowe, L. 1978. *Property Development*. London: E. and F.N. Spon Ltd

Campaign Against a Criminal Trespass Law 1978. *Whose Law and Order?* London: CACTL, c/o 35 Wellington Street, London WC2

Carey, L. and Mapes, R. 1972. *The Sociology of Planning: a Study of Social Activity on New Housing Estates*. London: Batsford Books

Castells, M. 1976. Theoretical propositions for an experimental study of urban social movements. In Pickvance (1976), pp. 147-73

————, 1977. *The Urban Question,* English edn. London: Arnold

Castle, B. 1976. *NHS Revisited*. London: Fabian Society, Fabian Tract 440

Centre for Environmental Studies 1975. *Proceedings of the Conference on Urban Change and Conflict*, Jan. 1975, at the University of York. London: Centre for Environmental Studies, Conference Paper 14

———, 1978. *Urban Change and Conflict Conference*, York, 1977. London: Centre for Environmental Studies, Conference Series 19

Chadwick, G.F. 1966. A systems view of planning. *Journal of the Town Planning Institute*, vol. 52, pp. 184-6

———, 1971. *A Systems View of Planning*. Oxford: Pergamon Press

Chapman, P. 1971. The Community Development Project. *Official Architecture and Planning*, vol. 34, no. 12, pp. 919-20

Clark, G. 1970. The lesson of Acklam Road. *New Statesman*, 7 Aug. 1970, pp. 139-40

———, 1972. Remember your humanity and forget the rest. In R. Benewick and T. Smith (eds.), *Direct Action and Democratic Politics*. London: Allen and Unwin, pp. 178-91

Clarke, S. 1977. Marxism, sociology and Poulantzas' theory of the state. *Capital and Class* (Bulletin of the Conference of Socialist Economists) (Summer), pp. 1-31

Clawson, M. and Hall, P. 1973. *Planning and Urban Growth: an Anglo-American Comparison*. Baltimore: Johns Hopkins University Press

Coates, K. and Silburn, R. 1972. The scope and limits of community action. *Community Action*, no. 1, pp. 15-17

Cockburn, C. 1970. *The Provision of Planning Education*. London: Centre for Environmental Studies, Information Paper 15

———, 1977. *The Local State*. London: Pluto Press

Cohen, P.S. 1968. *Modern Social Theory*. London: Heinemann

Colenutt, B. 1975. Behind the property lobby. In Conference for Socialist Economists (1975), pp. 123-33

Community Action 1972. Bureaucratic guerilla. *Community Action*, no. 1, pp. 26-8

———, 1973. Stirrings in Golborne. *Community Action*, no. 7, pp. 30-2

Community Development Project 1974. *Inter Project Report*. London: Community Development Project (CDP)

———, 1975. *Forward Plan 1975/76*. London: CDP Information and Intelligence Unit

———, 1976. *Whatever Happened to Council Housing?* London: CDP Information and Intelligence Unit

———, 1977a. *Gilding the Ghetto*. London: CDP Inter-project editorial team

_____, 1977b. *The Management of Deprivation*. London: Polytechnic of the South Bank

_____, 1977c. *The Poverty of the Improvement Programme*. Newcastle: CDP Political Economy Collective

Conference of Socialist Economists 1975. *Political Economy and the Housing Question*. London: Conference of Socialist Economists

_____, 1976. *Housing and Class in Britain*. London: Conference of Socialist Economists

Corrigan, P. and Leonard, P. 1978. *Social Work Practice under Capitalism: a Marxist Approach*. London and Basingstoke: Macmillan

Coser, L.A. and Rosenberg, B. 1964. *Sociological Theory: a Book of Readings*, 2nd edn. New York and London: Macmillan

Counter Information Services 1973. *The Recurrent Crisis of London. Anti-report on the Property Developers*. London: Counter Information Services (CIS)

_____, n.d. 1974? *Your Money and Your Life. Insurance Companies and Pension Funds*. London: CIS

Cowley, J., Kaye, A., Mayo, M. and Thompson, M. (eds.) 1977. *Community or Class Struggle?* London: Stage I

Creese, W.L. 1966. *The Search for Environment: the Garden City; Before and After*. New Haven: Yale University Press

Crenson, M.A. 1971. *The Unpolitics of Air Pollution: a Study of Non-Decision Making in the Cities*. Baltimore and London: Johns Hopkins Press

Crosby, T. 1973. *How to Play the Environment Game*. Harmondsworth: Penguin Books

Crosland, C.A.R. 1956. *The Future of Socialism*. London: Cape

_____, 1962. *The Conservative Enemy: a Programme of Radical Reform for the 1960s*. London: Cape

Crossman, R.H.S. 1975. *The Diaries of a Cabinet Minister,* Vol.1. London: Hamilton and Cape

_____, 1976. *The Diaries of a Cabinet Minister,* Vol. 2. London: Hamilton and Cape

_____, 1977. *The Diaries of a Cabinet Minister,* Vol. 3. London: Hamilton and Cape

Crouch, C. (ed.) 1977. *British Political Sociology Yearbook, Vol. 3, Participation*. London: Croom Helm

Cullingford, D., Flynn, P. and Webber, R. 1975. *Liverpool Social*

*Area Analysis: Interim Report*. London: Centre for Environmental Studies

Cullingworth, J.B. 1972. *Problems of an Urban Society. Vol. 2. The Social Context of Planning*. London: Allen and Unwin

———, 1976. *Town and Country Planning in Britain*, 6th edn. London: Allen and Unwin

Dahl, R.A. 1961. *Who Governs? Democracy and Power in an American City*. New Haven: Yale University Press

———, 1967. *Pluralist Democracy in the United States: Conflict and Consensus*. Chicago: Rand McNally

Dahrendorf, R. 1959. *Class and Class Conflict in Industrial Society*. London: Routledge and Kegan Paul

Damer, S. and Hague, C. 1971. Public participation in planning: a review. *Town Planning Review*, vol. 42, pp. 217-32

Darke, J. and Darke, R. 1970. *Health and Environment: High Flats*. London: Centre for Environmental Studies, University Working Paper 10

Darke, R. and Walker, R. (eds.) 1977. *Local Government and the Public*. London: Leonard Hill

Davidoff, P. 1965. Advocacy and pluralism in planning. *Journal of the American Institute of Planners*, vol. 31, no. 4, pp. 331-7

Davidoff, P. and Reiner, T.A. 1962. A choice theory of planning. *Journal of the American Institute of Planners*, vol. 28

Davies, J.G. 1972. *The Evangelical Bureaucrat*. London: Tavistock

Deakin, N. 1968. The politics of the Commonwealth Immigrants Bill. *Political Quarterly*, vol. XXXIX, no. 1, pp. 25-45

Dearlove, J. 1973. *The Politics of Policy in Local Government. The Making and Maintenance of Public Policy in the Royal Borough of Kensington*. London: Cambridge University Press

Dennis, N. 1968. The popularity of the neighbourhood community idea. In Pahl (ed.) (1968), pp. 74-92

———, 1970. *People and Planning. The Sociology of Housing in Sunderland*. London: Faber

———, 1972. *Public Participation and Planners' Blight*. London: Faber

Department of the Environment 1977. *Inner Area Studies: Liverpool, Birmingham, Lambeth*. London: HMSO

Diamond, D. and McLoughlin, J.B. (eds.) 1973. Education for planning. *Progress in Planning*, vol. 1, Part 1

Dobb, M. 1970. *Socialist Planning: Some Problems*. London: Lawrence and Wishart

Domhoff, G.W. 1978. *Who Really Rules?* New Brunswick, New Jersey: Transaction Books

Donnison, D.V. 1967. *The Government of Housing.* Harmondsworth: Penguin Books

———, 1973. Micro-politics of the city. In D.V. Donnison and D.E.C. Eversley (eds.), *London, Patterns and Problems*, pp.383-404. London: Heinemann

Dunleavy, P.J. 1976. An issue centred approach to the study of power. *Political Studies*, vol. XXIV, no. 4, pp. 423-34

———, 1977a. The politics of high rise housing in Britain: local communities tackle mass housing. Unpublished D Phil thesis, Nuffield College, Oxford

———, 1977b. Protest and quiescence in urban politics: a critique of some pluralist and structuralist myths. *International Journal of Urban and Regional Research,* vol. 1, no. 2, pp. 193-218

Dyckman, J.W. 1961. Planning and decision theory. *Journal of the American Institute of Planners*, vol. 27, no. 4, pp. 335-45

Eckstein, H.H. 1960. *Pressure Group Politics: the Case of the BMA.* London: Allen and Unwin

Elkin, S.L. 1974. *Politics and Land Use Planning. The London Experience.* London: Cambridge University Press

Elliott, B. and McCrone, D. 1975. Landlords as urban managers: a dissenting opinion. In Centre for Environmental Studies (1975), pp. 31-61

Elliott, D. 1977. *The Lucas Aerospace Workers' Campaign.* London: Fabian Society, Young Fabian Pamphlet 46

Etzioni, A. 1968. *The Active Society.* New York: Free Press

Eversley, D.E.C. 1972. Rising costs and static incomes: some economic consequences of regional planning in London. *Urban Studies*, vol. 9, no. 3, pp. 347-68

———, 1973. *The Planner in Society: the Changing Role of a Profession.* London: Faber

Falk, N. 1974. Community as the developer. *Built Environment*, vol. 3, no. 4, pp. 192-6

Falk, N. and Martinos, H. 1975. *Local Government and Economic Renewal.* London: Fabian Society, Fabian Research Series 320

Faludi, A. (ed.) 1973. *A Reader in Planning Theory.* Oxford: Pergamon Press

Ferris, J. 1972. *Participation in Urban Planning: the Barnsbury Case.* London: G. Bell and Sons

Feyerabend, P. 1975. *Against Method.* London: New Left Books

____, 1978. *Science in a Free Society*. London: New Left Books

Finer, S.E. 1955. The political power of private capital. *Sociological Review*, vol. 3, no. 2, pp. 279-94 (Part I)

____, 1956. The political power of private capital. *Sociological Review*, vol. 4, no. 1, pp. 5-30 (Part II)

____, 1958. *Anonymous Empire: a Study of the Lobby in Great Britain*. London: Pall Mall

Fletcher, C. 1976. The relevance of domestic property. *Sociology*, vol. 10, no. 3, pp. 451-68

Flynn, N. 1978. Urban experiments limited: lessons from CDP and the Inner Area Study. Centre for Environmental Studies (1978), pp. 33-77

Foley, D.L. 1960. British town planning: one ideology or three? *British Journal of Sociology*, vol. XI, no. 3, pp. 211-31

Forester, T. 1978a. The micro-electronic revolution. *New Society*, 9 Nov., pp. 330-2

____, 1978b. Society with chips and without jobs. *New Society*, 16 Nov., pp. 387-8

Fox, A. 1973. Industrial relations: a social critique of pluralist ideology. In J. Child (ed.), *Man and Organisation*, pp. 185-233. London: Allen and Unwin

Frankel, H. 1970. *Capitalist Society and Modern Sociology*. London: Lawrence and Wishart

Frankenberg, R. 1966. *Communities in Britain*. Harmondsworth: Penguin Books

Franks, M. 1974. Whose initiative? *Built Environment*, vol. 3, no. 11, p. 547

Fraser, D. 1973. *The Evolution of the British Welfare State*. London: Macmillan

Friedman, A.L. 1977. *Industry and Labour: Class Struggle at Work and Monopoly Capitalism*. London: Macmillan

Friedmann, W. 1974. *Public and Private Enterprise in Mixed Economies*. London: Stevens

Friend, J.K. and Jessop, W.N. 1969. *Local Government and Strategic Choice*. London: Tavistock

Friend, J.K., Power, J. and Yewlett, C. 1974. *Public Planning: the Inter Corporate Dimension*. London: Tavistock

Fry, G.K. 1969. *Statesmen in Disguise: the Changing Role of the Administrative Class of the British Home Civil Service, 1853-1966*. London: Macmillan

Galbraith, J.K. 1957. *American Capitalism: the Concept of*

*Countervailing Power*. London: Hamilton

———, 1967. *The New Industrial State*. London: Hamilton

———, 1970. *The Affluent Society*. Harmondsworth: Penguin Books

———, 1974. *Economics and the Public Purpose*. London: Deutsch

Gans, H.J. 1968. *People and Plans*. London: Basic Books

Garrett, J. 1972. *The Management of Government*. Harmondsworth: Penguin Books

Gavin, B. and O'Malley, J. 1977. What are our long-term political objectives? In Cowley *et al.* (1977), pp. 202-9

Geddes, P. 1949. *Cities in Evolution*. London: Williams and Norgate

de George, R.T. and Fernande, M. (eds.) 1972. *The Structuralists from Marx to Lévi-Strauss*. New York: Doubleday, Anchor Books

George, V. and Wilding, P. 1976. *Ideology and Social Welfare*. London: Routledge and Kegan Paul

Gerth, H.H. and Mills, C.W. 1946. *From Max Weber: Essays in Sociology*. Oxford: Oxford University Press

Giddens, A. 1973. *The Class Structure of Advanced Societies*. London: Hutchinson

Gitlin, T. 1969. Local pluralism as theory and ideology. In H.P. Dreitzel (ed.), *Recent Sociology no. 1,* pp. 62-87. London: Macmillan

Glass, R. 1955. Urban sociology in Great Britain: a trend report. *Current Sociology*, vol. 4, nos. 1 and 4, pp. 25-76

———, 1959. The evaluation of planning. *International Social Science Journal*, vol. 11, no. 3, pp. 393-409

———, 1970. Housing in Camden. *Town Planning Review*, vol. 41, no. 4, pp. 15-40

Glass, R. and Westergaard, J.H. 1965. *London's Housing Needs*. Statement of evidence to the committee on Greater London (Milner-Holland). London: Centre for Urban Studies, University College

Glyn, A. and Sutcliffe, B. 1971. The critical condition of British capital. *New Left Review,* no. 66, pp. 3-33

Goddard, D. 1972. Anthropology: the limits of functionalism. In Blackburn (ed.) (1972), pp. 61-75

Goldmann, L. 1969. *The Human Sciences and Philosophy*. London: Cape

Goodman, R. 1972. *After the Planners*. Harmondsworth: Penguin Books

Gorz, A. 1977. The reproduction of labour power: the model of consumption. In Cowley *et al.* (1977), pp. 22-39

Gough, I. 1975. State expenditure in advanced capitalism. *New Left Review,* no. 92, pp. 53-92

Gouldner, A.W. 1973. The sociologist as partisan: sociology and the welfare state. In *For Sociology: Renewal and Critique in Sociology Today,* pp. 27-68. London: Allen Lane

Gracey, H. 1969. *Sociology of Planning and Urban Growth.* London: Centre for Environmental Studies, University Working Paper 7

Greenwood, R. and Stewart, J.D. 1974. *Corporate Planning in English Local Government: an Analysis with Readings 1967-1972.* London and Tonbridge: Charles Knight

Greve, J., Page, D. and Greve, S. 1972. *Homelessness in London.* London: Chatto and Windus

Grove, J.L. and Procter, S.C. 1966. Citizen participation in planning. *Journal of the Town Planning Institute,* vol. 52, no. 10, pp. 414-16

Gutsman, W.L. (ed.) 1969. *The English Ruling Class.* London: Weidenfeld and Nicolson

Hall, C. 1974. *How to Run a Pressure Group.* London: Dent

Hall, P. 1972. Planning and the environment. In P.Townsend and N. Bosanquet (eds.), *Labour and Inequality,* pp. 261-73. London: Fabian Society

\_\_\_\_, 1973. *The Containment of Urban England,* 2 vols. London: Allen and Unwin

\_\_\_\_, 1975. *Urban and Regional Planning.* Newton Abbott: David and Charles

\_\_\_\_, 1977. *The World Cities,* 2nd edn. London: Weidenfeld and Nicolson

Halmos, P. (ed.) 1973. *Professionalisation and Social Change.* Sociological Review Monograph no. 20. Keele: University of Keele

Halsey, A.H. (ed.) 1972. *Educational Priority.* London: HMSO

Hanson, A.H. 1969. Public administration and the social order in twentieth-century Britain. In A.H. Hanson, *Planning and the Politicians,* pp. 91-103. London: Routledge and Kegan Paul

Harloe, M. (ed.) 1977. *Captive Cities.* London: John Wiley

Harris, D. 1966. The regional problem: why interfere? *Official Architecture and Planning,* vol. 29, no. 3, pp. 427-9

Harrison, M. 1960. *Trade Unions and the Labour Party since 1945.*

London: Allen and Unwin

Harrison, M.L. 1972. Development control — the influence of political, legal and ideological factors. *Town Planning Review,* vol. 43, no. 3, pp. 254-74

Harvey, D. 1973. *Social Justice and the City.* London: Arnold

——, 1978. Population, resources, and the ideology of science. In R. Peet (ed.), *Radical Geography: Alternative Viewpoints on Contemporary Social Issues,* pp. 213-42. London: Methuen

Hayward, J. and Watson, M. (eds.) 1975. *Planning, Politics and Public Policy: British, French and Italian Experience.* Cambridge: Cambridge University Press

Heap, D. 1973. *An Outline of Planning Law,* 6th edn. London: Sweet and Maxwell

Hill, D. 1970. *Participating in Local Affairs.* Harmondsworth: Penguin Books

Hill, M.J. 1972. *The Sociology of Public Administration.* London: Weidenfeld and Nicolson

Hillman, J. 1971. *Planning for London.* Harmondsworth: Penguin Books

Hindell, K. 1965. The genesis of the Race Relations Bill. *Political Quarterly,* vol. XXXVI, no. 4, pp. 390-405

Hindess, B. 1971. *The Decline of Working Class Politics.* London: MacGibbon and Kee

Hirsch, J. 1978. The state apparatus and social reproduction: elements of a theory of the bourgeois state. In J.Holloway and S.Picciotto (eds.) (1978), pp. 57-107

HMSO 1940. *Report of the Royal Commission on the Distribution of the Industrial Population* (Barlow Report). Cmnd 6153. London: HMSO

1942a. *Report of the Committee on Land Utilisation in Rural Areas* (Scott Report). Cmnd 6378. London: HMSO

1942b. *Report of the Expert Committee on Compensation and Betterment* (Uthwatt Report). Cmnd 6386. London: HMSO

1950. *Report of the Committee on the Qualifications of Planners* (Schuster Report). Cmnd 8059. London: HMSO

1963. *Traffic in Towns* (Buchanan Report). London: HMSO

1965. *The Future of Development Plans.* Report of the Planning Advisory Group. London: HMSO

1967a. *Report of the Committee on the Management of Local Government* (Maud Report), vol. 1. London: HMSO

1967b. *Report of the Committee on the Staffing of Local Govern-*

*ment* (Mallaby Report). London: HMSO

1968a. *Report of the Committee on the Civil Service* (Fulton Report). London: HMSO

1968b. *Report of the Committee on Local Authority and Allied Personal Social Services* (Seebohm Report). Cmnd 3703. London: HMSO

1969a. *The Intermediate Areas* (Hunt Report). Cmnd 3998. London: HMSO

1969b. *People and Planning.* Report of the Committee on Public Participation in Planning (Skeffington Report). London: HMSO

1970. *The Reorganisation of Central Government.* Cmnd 4506. London: HMSO

1972a. *Study Group on Local Authority Management Structure* (Bains Report). London: HMSO

1972b. *Report of the Working Party on Local Authority/Private Enterprise Schemes.* London: HMSO

1974. *Land,* White Paper. Cmnd 5730. London: HMSO

1975a. *Review of the Development Control System, Final Report* (Dobry Report). London: HMSO

1975b. *The Manning of the Public Services in London.* London: HMSO

1978. *New Programmes for the Unemployed.* LBW Paper 9394. Report by Director of Planning on Manpower Services Commission. London: HMSO

Hoinville, G. and Jowell, R. 1972. Will the real public please stand up? *Official Architecture and Planning,* vol. 35, no. 3, pp. 159-61

Holland, S. 1976. *Capital versus the Regions.* London: Macmillan

Holloway, J. and Picciotto, S. (eds.) 1978. *State and Capital. A Marxist Debate.* London: Arnold

Hounslow Hospital Occupation Committee, Elizabeth Garrett Anderson Joint Shop Stewards' Committee, Plaistow Maternity Action Committee, Save St Nicks Hospital Campaign 1978. *Keeping Hospitals Open.* London: the four hospital campaigns

Howard, E. 1902. *Garden Cities of Tomorrow,* being the second edition of *Tomorrow: a Peaceful Path to Real Reform.* London

Hunt, A. 1977. Theory and politics in the identification of the working class. In A.Hunt (ed.), *Class and Class Structure,* pp. 81-111. London: Lawrence and Wishart

Hunter, J. 1969. Self-help planning. *Architecture and Building*

*News*, 20 Nov., pp. 34-9

Illich, I. 1971. *Deschooling Society*. London: Calder and Boyars

———, 1976. *Celebration of Awareness: a Call for Institutional Revolution*. Harmondsworth: Penguin Books

———, 1977. *Limits to Medicine: Medical Nemesis, the Expropriation of Health*. Harmondsworth: Penguin Books

———, 1978. *The Right to Useful Unemployment — and its Professional Enemies*. London: Boyars

James, S. and Costa, M.D. 1972. *The Power of Women and the Subversion of the Community*, 2nd English edn. Bristol: Falling Wall Press

Jay, A. 1972. *The Householders' Guide to Community Defence against Bureaucratic Aggression*. London: Cape

Jeffery, N. and Caldwell, M. (eds.) 1977. Planning and urbanisation in China. *Progress in Planning*, vol. 8, Part 2

Jephcott, P. and Robinson, H. 1971. *Homes in High Flats*. Edinburgh: Oliver and Boyd

Jerman, B. 1971. *Do Something!: a Guide to Self-help Organisations*. London: Garnstone Press

Johnson, T. 1972. *Professions and Power*. London: Macmillan

Johnson, T.A. 1974. *Calculations for Development Gains Tax*. Reading: Centre of Advanced Land Use Studies, College of Estate Management

Jones, D. and Mayo, M. (eds.) 1974. *Community Work: One*. London: Routledge and Kegan Paul

———, 1975. *Community Work: Two*. London: Routledge and Kegan Paul

*Journal of the Town Planning Institute* 1972. 'Pragma' in *Journal of the Town Planning Institute*, vol. 58, no. 8, p. 340

Jowell, J. 1977. Bargaining in development control. *Journal of Planning Law* (June), pp. 414-33

Kahn, H., Brown, W. and Martel, L. 1978. *The Next 200 years*. London: Sphere Books

Kapp, K.W. 1978. *The Social Costs of Business Enterprise*. Nottingham: Spokesman Books

Kaye, A. and Thompson, M. 1977. The class basis of planning. In Cowley *et al.* (1977), pp. 101-7

Keeble, L.B. 1966. The role of the councillor in town planning. *Journal of the Town Planning Institute*, vol. 52, no. 6, pp. 219-22

Kennett, M.J.C. 1968. Planners and the public. *Official Architecture and Planning*, vol. 31, no. 4, p. 549

Keynes, J.M. 1926. *The End of Laissez Faire*. London: Hogarth
____, 1936. *The General Theory of Employment, Interest and Money*. London: Macmillan
Kimber, R. and Richardson, J.T. 1974. *Campaigning for the Environment*. London: Routledge and Kegan Paul
Kingham, M. 1977. *Squatters in London*. London: Shelter
Knowles, R.S.B. 1977. *Modern Management in Local Government*, 2nd edn. London: Barry Rose
Konrad, G. and Szelenyi, I. 1977. Social conflicts of under urbanisation. In Harloe (ed.) (1977), pp. 157-73
Kovačerić, Z. 1972. Belgrade. In Robson and Regan (eds.) (1972), pp. 165-98
Kuhn, T.S. 1962. *The Structure of Scientific Revolutions*. Chicago: University of Chicago Press
Lakatos, I. 1970. Falsification and the methodology of scientific research programmes. In I. Lakatos and A. Musgrave (eds.), *Criticism and the Growth of Knowledge*, pp. 91-196. Cambridge: Cambridge University Press
Lamarche, F. 1976. Property development and the economic foundations of the urban question. In Pickvance (ed.) (1976), pp. 85-118
Lambert, J.R. 1975. Housing class and community action in a redevelopment area. In C. Lambert and D. Weir (eds.), *Cities in Modern Britain*, pp. 415-24. Glasgow: Fontana/Collins
Lambert, J.R., Paris, C. and Blackaby, B. 1978. *Housing Policy and the State. Allocation, Access and Control*. London: Macmillan
Land Campaign Working Party 1975. *Lie of the Land. Community Land Act. Land Nationalisation Betrayed*. London: Land Campaign Working Party, c/o 31 Clerkenwell Close, London EC1.
Lane, M. (ed.) 1970. *Structuralism: a Reader*. London: Cape
Lapping, B. and Raddice, G. 1968. *More Power to the People*. London: Longmans
Leech, A. 1971. A plan from the people. *Surveyor*, 26 Feb., pp. 24-5
Lefebvre, H. 1976. The politics of space. *Antipode*, vol. 8, no. 2, pp. 30-7
Levin, P. 1968. Do the institutes really want public participation? *Journal of the Royal Institute of British Architects*, vol. 75, no. 11, pp. 495-6
____, 1969. The planning inquiry farce. *New Society*, 3 July, pp. 17-18

———, 1971a. The amenity movement. *Official Architecture and Planning,* vol. 34, no. 11, pp. 846-9

———, 1971b. Participation: the planners *v.* the public. *New Society,* 24 June, pp. 1090-1

———, 1976. *Government and the Planning Process.* London: Allen and Unwin

Levin, P. and Donnison, D. 1969. People and planning. *Public Administration,* vol. 47 (Winter), pp. 473-9

Lewis, J. 1977. British capitalism, the welfare state and the first radicalisation of state employees. In Cowley *et al.* (1977), pp. 112-24

Lichfield, N. 1960. Cost-benefit analysis in city planning. *Journal of the American Institute of Planners,* vol. 26, no. 4, pp. 273-9

———, 1968. Goals in planning. In *Report of the Proceedings. Town Planning Institute Summer School.* Manchester: Town Planning Institute

Lindberg, L.N., Alford, R., Crouch, C. and Offe, C. (eds.) 1975. *Stress and Contradiction in Modern Capitalism.* London: Lexington Books

Lindblom, C.E. 1959. The science of 'muddling through'. *Public Administration Review* (Spring), pp. 79-88

Lipset, S.M. 1960. *Political Man: the Social Bases of Politics.* New York: Garden City Press

Lipsky, M. 1970. *Protest in City Politics. Rent Strikes, Housing and the Power of the Poor.* Chicago: Rand McNally

Lojkine, J. 1976. Contribution to a Marxist theory of capitalist urbanization. In Pickvance (ed.) (1976), pp. 119-46

———, 1977. Big firm's strategies, urban policy and urban social movements. In Harloe (ed.) (1977), pp. 141-56

Loughlin, M. 1978. Bargaining as a tool of development control. A case of all gain and no loss? *Journal of Planning and Environment Law,* pp. 290-5

Lukes, S. 1974. *Power: a Radical View.* London: Macmillan

McConaghy, D. 1971. Inner Area Agencies. *Official Architecture and Planning,* vol. 34, no. 5, pp. 353-6

———, 1972. The limitations of advocacy. *Journal of the Royal Institute of British Architects,* vol. 79, no. 2, pp. 63-6

McCulloch, J. 1978. *Meanwhile Gardens.* London: Gulbenkian Foundation

McDougall, G. 1973. The systems view of planning: a critique. *Socio-Economic Planning Science,* vol. 7, pp. 79-90

McKean, C. 1977. *Fight Blight: a Practical Guide to the Causes of Urban Dereliction and What People Can Do about It.* London: Kaye and Ward

McKenzie, R.T. 1958. Parties, pressure groups and the British political process. *Political Quarterly* (Jan. - March), pp. 5-16

——, 1964. *British Political Parties.* London: Heinemann

McKeown, T. 1965. *Medicine in Modern Society.* London: Allen and Unwin

McLoughlin, J.B. 1969. *Urban and Regional Planning — a Systems Approach.* London: Faber

Marcus, S. 1969. Planners — who are you? *Journal of the Town Planning Institute,* vol. 57, pp. 54-9

Marriott, O. 1967. *The Property Boom.* London: Pan Books

Marris, P. and Rein, M. 1967. *Dilemmas of Social Reform.* Harmondsworth: Penguin Books

Marx, K. 1954. *Capital,* Vol. I. London: Lawrence and Wishart

Mason, T. 1978. Community action and the local authority: a study in the incorporation of protest. In Centre for Environmental Studies (1978), pp. 89-116

Massey, D. and Catalano, A. 1978. *Capital and Land: Landownership by Capital in Great Britain.* London: Arnold

Mazziotti, D.F. 1974. The underlying assumptions of advocacy planning: pluralism and reform. *Journal of the American Institute of Planners* (Jan.), pp. 38-48

Meadows, D.H., Meadows, D.L., Randers, J. and Behrens, W.W. 1974. *The Limits to Growth.* London: Pan Books

Mellor, R. 1975. Urban sociology in an urbanised society. *British Journal of Sociology,* vol. XXVI, no. 3, pp. 276-93

——, 1977. *Urban Sociology in an Urbanized Society.* London: Routledge and Kegan Paul

Merton, R.K. 1949. *Social Theory and Social Structure.* New York: Free Press

Miliband, R. 1969. *The State in Capitalist Society.* London: Weidenfeld and Nicolson

——, 1970. The capitalist state: reply to Nicos Poulantzas. *New Left Review,* no. 59, pp. 53-60

——, 1973. The power of labour and the capitalist enterprise. In J. Urry and J. Wakeford (eds.), *Power in Britain,* pp. 136-45. London: Heinemann

Mills, C.W. 1956. *The Power Elite.* New York and London: Oxford University Press

\_\_\_\_, 1959. *The Causes of World War Three*. London: Secker and Warburg

Mingione, E. 1977. Theoretical elements for a Marxist analysis of urban development. In Harloe (ed.) (1977), pp. 89-109

Minkin, L. 1977. The Labour Party has not been hijacked. *New Society,* 6 Oct., pp. 6-8

Minns, R. and Thornley, J. 1978. *State Shareholding. The Role of Local and Regional Authorities.* London and Basingstoke: Macmillan

Mishan, E.J. 1967. *The Costs of Economic Growth.* London: Staples Press

Mitchell, J. 1971. *Woman's Estate.* Harmondsworth: Penguin Books

Muchnick, D.M. 1970. *Urban Renewal in Liverpool.* London: G. Bell and Sons

Murie, A.S., Niner, P. and Watson, C. 1976. *Housing Policy and the Housing System.* London: Allen and Unwin

Musil, J. 1968. The development of Prague's ecological structure. In Pahl (ed.) (1968), pp. 232-59

Muth, R.F. 1968. *Cities and Housing: the Spatial Pattern of Urban Residential Land Use.* Chicago: Chicago University Press

Newby, H., Bell, C., Rose, D. and Saunders, P. 1978. *Property, Paternalism and Power. Class and Control in Rural England.* London: Hutchinson

Norman, P. 1975. Managerialism: review of recent work. In Centre for Environmental Studies (1975), pp. 62-86

O'Connor, J. 1973. *The Fiscal Crisis of the State.* New York: St. Martin's Press

Olives, J. 1976. The struggle against urban renewal in the Cité d'Aliarte (Paris). In Pickvance (ed.) (1976), pp. 174-97

Osborn, F.J. and Whittick, A. 1969. *The New Towns: the Answer to Megalopolis.* London: Leonard Hill

Pahl, R.E.(ed.) 1968. *Readings in Urban Sociology.* Oxford: Pergamon Press

\_\_\_\_, 1970. *Patterns of Urban Life.* London: Longmans

\_\_\_\_, 1975. *Whose City?* 2nd edn. Harmondsworth: Penguin Books

\_\_\_\_, 1977. 'Collective consumption' and the state in capitalist and state socialist societies. In Scase (ed.) (1977), pp. 153-71

\_\_\_\_, 1978. 'Will the inner city problem ever go away?' *New Society,* 28 Sept., pp. 678-81.

Pahl, R.E. and Winkler, J.T. 1974. The coming corporatism. *New*

*Society,* 10 Oct., pp. 72-6

Palmer, J.A.D. 1972. Introduction to the British edition. In Goodman (1972), pp. 9-50

———, 1974. The aims of social planning and the uses of information. In L.Grayson (ed.), *Social Planning: Sources of Information,* pp. 8-18. London: Library Association

Parenti, M. 1970. Power and pluralism: a view from the bottom. *Journal of Politics,* vol. 32, pp. 501-30

Paris, C. 1977. Housing Action Areas. *Roof* (Jan.), pp. 9-14

Parsons, T. 1949. *The Structure of Social Action.* New York: Free Press

———, 1951. *The Social System.* New York: Free Press

Pearce, D.W.(ed.) 1978. *The Valuation of Social Cost.* London: Allen and Unwin

Peattie, L.R. 1968. Reflections on advocacy planning. *Journal of the American Institute of Planners,* vol. 34, no. 2, pp. 80-8

Perman, D. 1973. *Cublington.* London: Bodley Head

Peterson, W. 1968. The ideological origins of Britain's New Towns. *Journal of the American Institute of Planners,* vol. 34, no. 3, pp. 160-70

Pickvance, C.G. 1974. On a materialist critique of urban sociology. *Sociological Review,* vol. 22, no. 2, pp. 203-20

———, 1975. From social base to social force. Some analytical issues in the study of urban protest. In Centre for Environmental Studies (1975), pp. 200-18

———, (ed.) 1976. *Urban Sociology: Critical Essays.* London: Tavistock

———, 1977. Marxist approaches to the study of urban politics: divergences among some recent French studies. *International Journal for Urban and Regional Research,* vol. 1, no. 2, pp. 219-55

———, 1978. *New Directions in Urban Sociology.* London: Tavistock

Pinker, R. 1968. The contribution of the social scientist in positive discrimination programmes. *Social and Economic Administration,* vol. 2, pp. 227-41

Playford, J. 1971. The myth of pluralism. In F.G. Castles, D.J. Murray and D.C. Potter (eds.), *Decisions, Organisations and Society.* Harmondsworth: Penguin Books/Open University Press

Polsby, N.W. 1963. *Community Power and Political Theory.* New Haven: Yale University Press

Popper, K.R. 1959. *The Logic of Scientific Discovery.* London: Hutchinson

____, 1963. *Conjectures and Refutations: the Growth of Scientific Knowledge.* London: Routledge and Kegan Paul

Potter, A.M. 1961. *Organised Groups in British National Politics.* London: Faber

Poulantzas, N. 1969. The problem of the capitalist state. *New Left Review,* no. 58, pp. 67-78

____, 1973. *Political Power and Social Classes.* London: New Left Books

Pym, B. 1973. The making of a successful pressure group. *British Journal of Sociology,* vol. XXIV, no. 4, pp. 448-61

Raban, J. 1975. *Soft City.* Glasgow: Fontana/Collins

Ratcliffe, J. 1974. Planning gain is not the answer. *Built Environment,* vol. 3, no. 3, pp. 148-9

Redpath, R. 1973. *Public Participation in Swinbrook: a New Model?* Unpublished GLC research paper, mimeo.

Rein, M. 1969. Social planning: the search for legitimacy. *Journal of the American Institute of Planners,* vol. XXXV, no. 4, pp. 233-44

Rex, J. 1961. *Key Problems of Sociological Theory.* London: Routledge and Kegan Paul

Rex, J. and Moore, R. 1967. *Race, Community and Conflict: a Study of Sparkbrook.* London: Oxford University Press

Reynolds, J.P. (ed.) 1969. Public participation in planning. *Town Planning Review,* vol. 40, no. 2, pp. 131-48

Richards, P.G. 1975. *The Reformed Local Government System.* London: Allen and Unwin

Richter, I. 1973. *Political Purpose in Trade Unions.* London: Allen and Unwin

Roberts, B.C. 1956. *Trade Union Government and Administration in Great Britain.* London: London School of Economics and Political Science

Roberts, M. 1974. *An Introduction to Town Planning Techniques.* London: Hutchinson

Robertson, J. 1978. *The Sane Alternative.* London: James Robertson, 7 St. Ann's Villas, London Wll

Robey, D. (ed.) 1973. *Structuralism: an Introduction.* Oxford: Clarendon Press

Robson, W.A. and Regan, D.E. (eds.) 1972. *Great Cities of the*

*World. Their Government, Politics and Planning,* 2 vols., 3rd edn. London: Allen and Unwin

Rock, D. 1972. The architecture of compromise. *Built Environment* (Oct.), pp. 448-53

Rose, H. 1973. Up against the Welfare State. In R. Miliband and J. Saville (eds.), *Socialist Register,* pp. 179-204. London: Merlin Press

Rose, H. and Hanmer, J. 1975. Community participation and social change. In Jones and Mayo (1975), pp. 25-45

Rose, H. and Puckett, T. 1973. Blueprint for bureaucrats. *Journal of the Royal Institute of British Architects,* vol. 80, pp. 277-81

Rose, H. and Rose, S. (eds.) 1976a. *The Political Economy of Science.* London and Basingstoke: Macmillan

____, 1976b. *The Radicalisation of Science.* London and Basingstoke: Macmillan

Roszak, T. 1971. *The Making of a Counter-culture: Reflections on the Technocratic Society and its Youthful Opposition.* London: Faber

____, 1972. *Where the Wasteland Ends: Politics and Transcendence in Post-industrial Society.* London: Faber

Royal Institute of British Architects 1968. Evidence to Skeffington Committee. *Journal of the Royal Institute of British Architects,* vol. 75, no. 10, p. 402

____, 1970. People and planning. *Journal of the Royal Institute of British Architects.* vol. 77, no. 1, pp. 36-7

Royal Institute of Chartered Surveyors 1947. *Memorandum of Observations on Town and Country Planning Bill* (Feb./March). London: RICS

____, 1974. What next for property? *Investors' Review,* vol. 82, no. 15, p. 8

Royal Town Planning Institute 1968. Public participation in planning. Memorandum of evidence submitted by the Institute to the Skeffington Committee. *Journal of the Town Planning Institute,* vol. 54, no. 8, pp. 343-4

____, 1971a. *Examinations Handbook.* London: RTPI

____, 1971b. Forum on public participation in planning. Sessional meeting at the Town Planning Institute. *Journal of the Town Planning Institute,* vol. 57, no. 4, pp. 171-5

Russell, B. 1935. *In Praise of Idleness and Other Essays.* London: Allen and Unwin

Saunders, P. 1975. They make the rules. *Policy and Politics,* no. 4, pp. 31-58

———, 1978. Domestic property and social class. *International Journal of Urban and Regional Research,* vol. 2, no. 2, pp. 233-51

Scase, R. (ed.) 1977. *Industrial Society: Class, Cleavage and Control.* London: British Sociological Association, Allen and Unwin

Schumpeter, J.A. 1943. *Capitalism, Socialism and Democracy.* London: Allen and Unwin

Selekman, S.K. and Selekman, B.M. 1956. *Power and Morality in a Business Society.* New York: McGraw-Hill

Selznick, P. 1949. *TVA and the Grassroots: a Study in the Sociology of Informal Organisations.* Berkeley: University of California Press

Shelter Community Action Team n.d. *The Great Sales Robbery.* London: SCAT, 31 Clerkenwell Close, London EC1

Shipley, P. (ed.) 1976. *Guardian Directory of Pressure Groups and Representative Associations.* London: Wilton House

Simmie, J. 1974. *Citizens in Conflict: the Sociology of Planning.* London: Hutchinson

Sklair, L. 1975. The struggle against the Housing Finance Act. In R. Miliband and J. Saville (eds.), *Socialist Register,* pp. 250-92. London: Merlin Press

Stewart, J.D., Armstrong, R.H.R. and Eddison, T. 1968. *Management Methods and Approaches for Planning.* Town and Country Planning Summer School. London: Town Planning Institute

Stewart, T. 1973. *Social Planning: an Overview.* Falmer, Sussex: University of Sussex, Planning and Transport Research and Computation

Stone, P.A. 1972. The economics of the form and organisation of cities. *Urban Studies,* vol. 9, no. 3, pp. 329-46

Strachey, J. 1956. *Contemporary Capitalism.* London: Gollancz

Stringer, P. and Taylor, M. 1972. Plans as seen by the public. *Built Environment,* vol. 1, no. 6, pp. 405-8

Swann, J. 1975. The political economy of residential redevelopment in London. In Conference of Socialist Economists (1975), pp. 104-15

Tawney, R.H. 1961. *The Acquisitive Society.* Glasgow: Fontana/Collins

Taylor, R. 1978. Work sharing and worklessness. *New Society,* no. 23, pp. 452-4

Telling, A.E. 1973. *Planning Law and Procedure,* 4th edn. London: Butterworths

Thompson, J.M. 1977. The London motorway plan. In W. R. D. Sewell and J. T. Coppock (eds.), *Public Participation in Planning,* pp. 59-69. London: John Wiley

Thorburn, A. 1970. *London under Stress.* London: Town and Country Planning Association

Thornley, A. 1977. Theoretical perspectives on planning participation. *Progress in Planning,* vol. 7, Part 1

Titmuss, R.M. 1968. *Commitment to Welfare.* London: Allen and Unwin

Town and Country Planning Association 1974. Quoted in 'Planners bypass public', *Guardian,* 10 June

Tyme, J. 1978. *Motorways versus Democracy.* London: Macmillan

Volmer, H.W. and Mills, D.L. (eds.) 1966. *Professionalization.* Englewood Cliffs, New Jersey: Prentice-Hall

Walters, S.Z. 1972. *Is There Any Benefit in Cost-benefit Analysis?* London: GLC Intelligence Unit, Quarterly Bulletin 21

Wates, N. 1976. *The Battle for Tolmers Square.* London: Routledge and Kegan Paul

Webb, S. 1889. Economic factors. In G. B.Shaw (ed.), *Fabian Essays in Socialism.* London: Allen and Unwin

Webber, M. 1968. Planning in an environment of change. *Town Planning Review,* vol. 39, no. 3, pp. 179-95 and 277-95

Westergaard, J.H. 1972. Sociology: the myth of classlessness. In Blackburn (ed.) (1972), pp. 119-63

_____, 1977. Class inequality and 'corporatism'. In Hunt (ed.) (1977), pp. 165-86

Westergaard, J. and Resler, H. 1975. *Class in a Capitalist Society.* London: Heinemann Educational

Wilcox, D. and Richards, D. 1977. *London, the Heartless City.* London: Thames Television

Willer, J. 1971. *The Social Determination of Knowledge.* Englewood Cliffs, New Jersey: Prentice-Hall

Williams, J., Anderson, J. and Goddard, J. 1972. *Central Area Housing Study.* London: City of Westminster, Development Plan Research Report R1

Winkler, J.T. 1977. The corporate economy: theory and administration. In Scase (ed.) (1977), pp. 43-58

Wolff, R.P. 1965. Beyond tolerance. In R. P. Wolff, B. Moore Jr. and H. Marcuse, *A Critique of Pure Tolerance,* pp. 11-61. London: Cape

Wood, E.W., Brower, S.N. and Latimer, M.W. 1966. The planner's people. *Journal of the American Institute of Planners,* vol. 32, pp. 228-33

Wright, E.O. 1978. *Class, Crisis and the State.* London: New Left Books

Yaffe, D. 1973. The crisis of profitability: a critique of the Glyn-Sutcliffe thesis. *New Left Review,* no. 80, pp. 45-62

Zaretsky, E. 1976. *Capitalism, the Family and Personal Life.* London: Pluto Press

Zawadzki, S. 1972. Warsaw. In Robson and Regan (eds.) (1972), pp. 1041-75

Zweig, F. 1961. *The Worker in an Affluent Society: Family Life and Industry.* London: Heinemann

# NAME INDEX

Allaby, M. 127, 128, 190, 195
Allen, V. 52, 176
Althusser, L. 80, 88, 94n6, 129n11, 147
Ambrose, P. 38, 39-40, 48, 77, 189, 190, 197
Amos, F.J.C. 37, 53
Arblaster, A. 155
Arnstein, S.R. 156
Ashworth, W. 51

Bachrach, P. 60, 65
Bailey, R. 29, 115, 143, 159
Balibar, E. 80
Baran, P.A. 52
Baratz, M.S. 60, 65
Beckerman, W. 31, 127
Berry, F. 29
Beveridge, W.H. 121, 186
Blackburn, R. 21, 24, 25, 93, 190
Blair, T. 10, 76, 140, 160
Broadbent, T.A. 21, 24, 52, 57, 79, 93, 100, 104, 124, 125, 133, 134, 137, 191

Campaign against a Criminal Trespass Law 180
Castells, M. 10, 28, 78, 80, 82, 84, 86, 87, 88, 91, 92, 118, 120, 121, 122, 123, 125, 147, 156, 162, 171, 172, 174, 175, 183, 188, 189
Centre for Environmental Studies 79, 129, 171
Chadwick, G.F. 95, 101, 102, 103
Clark, G. 77, 160
Coates, K. 162
Cockburn, C. 107, 108, 109, 114, 133, 151, 152, 159
Colenutt, B. 38, 39-40, 48, 61, 77
Conference of Socialist Economists 52, 53, 79
Counter Information Services 48, 53, 54, 80, 124, 144, 188
Crosland, C.A.R. 20, 57, 116
Crossman, R.H.S. 34, 169

Cullingworth, J.B. 40, 51

Dahl, R.A. 10, 57, 63, 64, 110
Davies, J.G. 10, 59, 66, 70, 71, 72, 135, 137, 168
Dearlove, J. 60, 65, 120, 168, 169, 170, 175
Dennis, N. 10, 36, 57, 59, 66, 70, 71, 72, 93, 99, 112, 113, 117, 135, 137, 138, 156, 158, 160, 163, 165, 168, 178
Department of the Environment 15, 30, 76, 139, 145, 180n2
Donnison, D.V. 10, 29, 75, 76, 77, 116, 117, 149, 159, 166, 178
Dunleavy, P.J. 58, 63, 64, 65, 87, 88, 89, 92, 93, 120, 121, 125, 171, 175, 178

Elliott, B. 89, 123, 124
Eversley, D.E.C. 10, 37, 70, 76, 132, 133, 137, 146, 188, 189

Ferris, J. 59, 60, 75, 120, 169, 170
Foley, D.L. 131
Fox, A. 57, 93
Frankel, H. 21, 116
Friedman, A.L. 25, 79, 93, 114
Friend, J.K. 95

Galbraith, J.K. 20, 57, 121, 186
Gans, H.J. 17, 143
Glass, R. 51, 53, 75, 94, 188
Glyn, A. 52
Goodman, R. 75, 129, 132, 140, 151
Gorz, A. 190
Gough, I. 25
Gouldner, A.W. 70
Greve, J. 29

Hall, P. 30, 33
Halsey A.H. 12, 18, 77
Hanmer, J. 93, 159
Harloe, M. 79
Harvey D. 10, 12, 17, 75, 98, 145, 146

221

# SUBJECT INDEX

Milton Keynes UK
Ingram Content Group UK Ltd.
UKHW031148141024
449569UK00024B/961